ORCHIDS OF NIGERIA

ORCHIDS OF NIGERIA

Description of 104 species with 170 photographs
Including a check-list of Nigerian orchids
by C.Z. Tang & P.J. Cribb

L.B.SEGERBÄCK

CRC Press
Taylor & Francis Group
Boca Raton London New York

CRC Press is an imprint of the
Taylor & Francis Group, an **informa** business
A BALKEMA BOOK

Published by:
CRC Press/Balkema
Schipholweg 107C, 2316 XC Leiden, The Netherlands

ISBN 13: 978-90-6191-217-0 (hbk)

Visit the Taylor & Francis Web site at
http://www.taylorandfrancis.com

and the CRC Press Web site at
http://www.crcpress.com

Contents

Foreword

The present handbook is chiefly intended as an introduction to the little known orchid flora of Nigeria and to stimulate an interest in it. The work is by no means complete, there are even entire genera left out, mainly for the lack of pictorial material. But as the author has had the privilege to work and to make collections in rain forest country as well as in the savanna (in Imo State and Kaduna State, respectively), he hopes to have included a selection of species representative of Nigeria as a whole.

The geographical distribution of the species has been indicated by means of a rough division of the country into a northern 'N' and a southern part 'S', see figure 2. An (N) or (S) within brackets indicates that the plant is known to occur sporadically, under special conditions, in the north or in the south, respectively. Whenever the author has found a species to be locally abundant, this has been mentioned in the text. Apart from this, no further comments have been possible on the distribution of the plants.

The habitat, locality and time of flowering are given for the specimens used in the illustrations, but these data must not be considered as typical in any way.

For practical reasons, the Nigerian orchids have been divided into two main groups, terrestrials and epiphytes, in this book. The author believes that this arrangement might facilitate the identification of the genera, but it must be pointed out that there is little botanical justification for such a division.

Wherever feasible, data pertaining to floral parts have been uniformly tabulated to facilitate the identification.

For keys to the genera and species, the interested reader is referred to the *Flora of West Tropical Africa* (FWTA), Vol. III:1, see the bibliography on page 110. This gives the most thorough treatment available on West African orchids, and it is highly recommended in spite of the rather demanding scientific vocabulary used. The very simplified keys included in the present handbook are based on easily observed field characteristics, and refer *only* to the genera here described.

A glossary with explanations of botanical terms, especially such as are used in dealing with orchids, has been included in this handbook. It is hoped that it will also be of some use to the reader who wants to consult other, more specialized works on orchids.

The author would like to express his sincere thanks to Dr P.J.Cribb and his colleagues at the Royal Botanic Gardens, Kew, England, for their kind help with the identification of many of the species here described.

Sösdala, Sweden, May 1983 Börje Segerbäck

CHAPTER ONE

General characteristics

The orchids, or *Orchidaceae,* form a very large family of plants. It has the largest number of species – about 25 000 are known – and is represented in all the continents except Antarctica. As the members are found all over the world, growing in widely different climates, considerable diversity is to be expected within the family.

The photograph in figure 1 shows a small *Leucorchis* which can be found on the coast of Greenland in the Arctic Ocean. Figure 3 is of a large African *Angraecum.* Despite the many apparent differences, these two plants have certain, easily recognizable features in common – and in common with all other orchids, see figure 4 which shows the variation in shape in some African orchids.

THE FLORAL PARTS

The orchid perianth consists of six segments, arranged in two whorls of three – the outer sepals and the inner petals. In some African genera, such as *Cyrtorchis,* figure 4f, all the segments can look the same. In such a case, the term *tepal* is used to denote both sepals and petals. However in general one segment of the inner whorl (the petals) is very different from the others.

Fig. 1. *Leucorchis albida.*

This segment is called the *lip* or *labellum.* It is often brightly coloured, spotted, or provided with streaks, ridges, hairs or frills. The rear part of the lip is sometimes drawn out to form an often nectar-filled *spur,* as in figure 4f and 4h. This can be quite long in some species – 20 cm or more. The segment of the outer whorl opposite the lip, the *dorsal sepal,* is also frequently different in size and shape from the paired *lateral sepals.*

Another distinguishing feature of orchid flowers is the special organ called the *column.* This is unique, being the union of the male and female reproductive organs, see figure 5.

At the tip of the column there is usually the *anther,* the male reproductive organ. This often has an *anther cap* or *operculum,* a little lid which, when pushed aside or lifted, reveals the pollen, see figure 6. In orchids, the pollen grains are generally stuck together to form a discrete number of club-like masses known as *pollinia.* Each pollinium is attached to the *rostellum* by a stalk (the *caudicle* or *stipes*) which ends in a sticky disc, the *viscidium.*

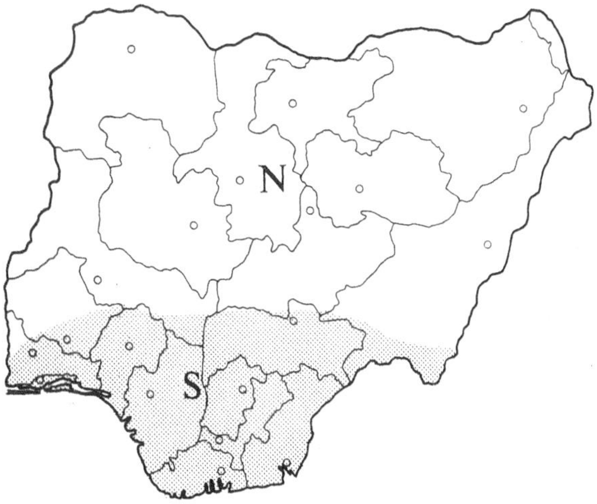

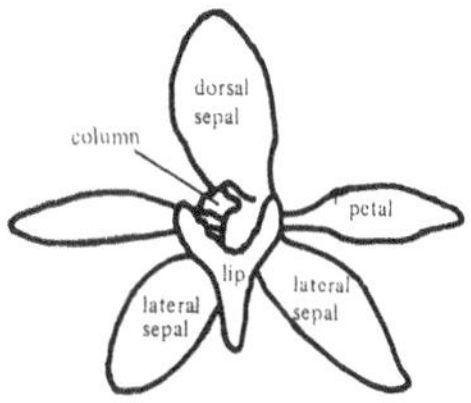

a) Ancistrochilus sp.

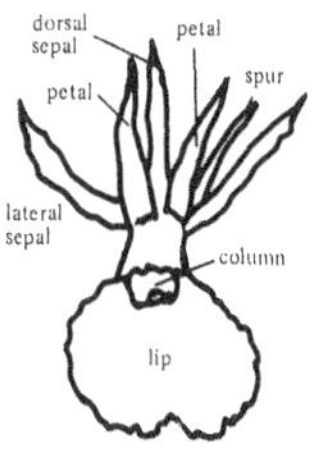

b) Eulophia sp.

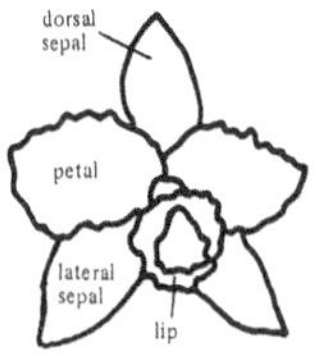

c) Vanilla sp.

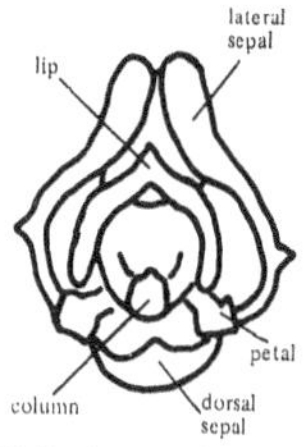

d) Polystachya sp.

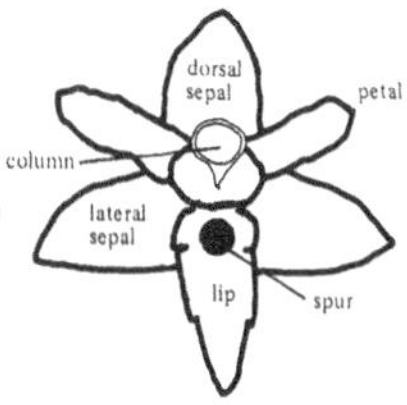

e) Tridactyle sp.

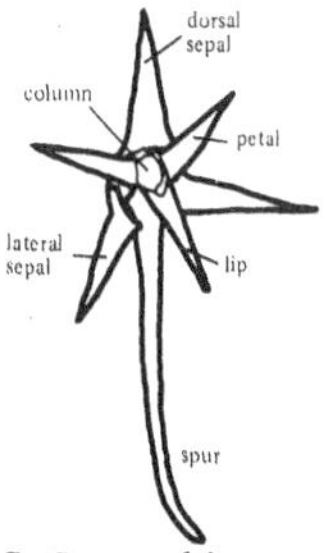

f) Cyrtorchis sp.

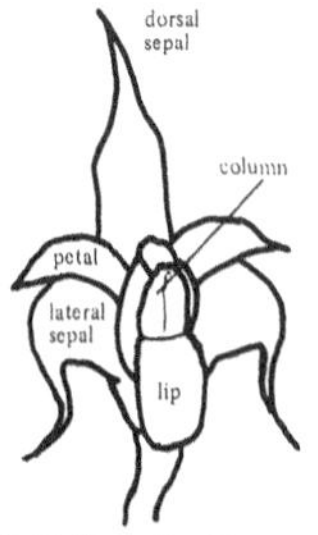

g) Bulbophyllum sp.

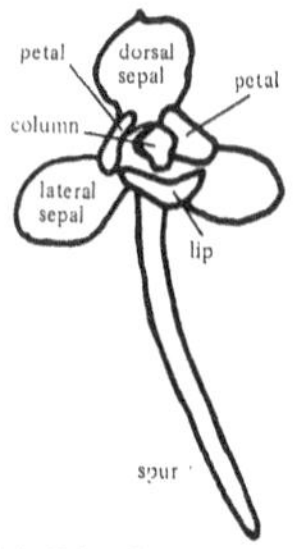

h) Diaphananthe sp.

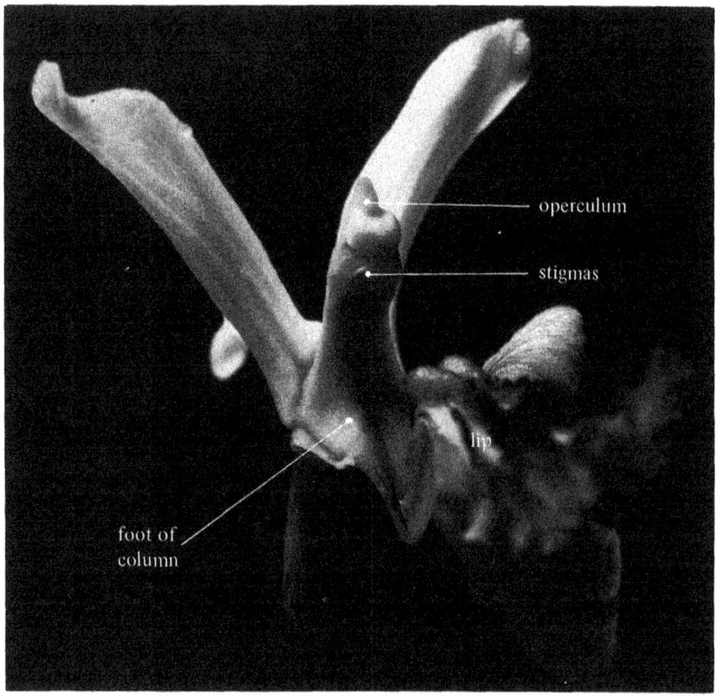

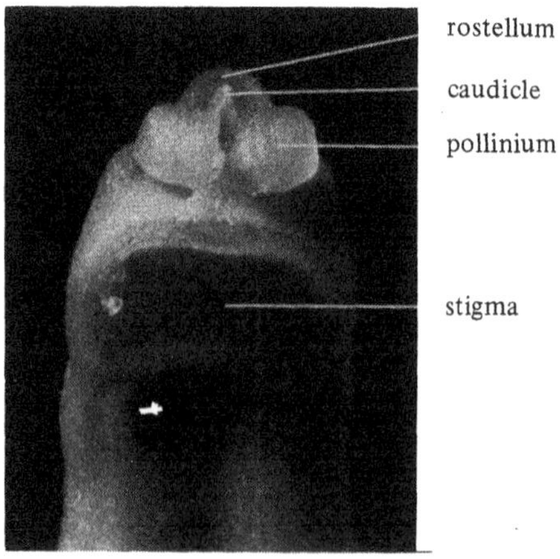

REPRODUCTION IN ORCHIDS

Pollination is by insects.* When one works its way into the flower to find nectar, the viscidium sticks on to its head or body. In this way, a whole pollinium will be transferred to the sticky stigma of another flower.

Generally there are two stigmas. These are found on the column below the anther, separated from it by the rostellum. (In fact this is a third stigma, but a sterile one.)

The fruit is nearly always a capsule. In most species, as the flower develops, the ovary turns 180° round its axis, a phenomenon known as *resupination.* This has the result that in these species, the lip comes to lie at the bottom of the flower. An important exception is the genus *Polystachya,* see pages 47 to 57.

The seeds are very small and very many. Orchid seeds are normally spread by the wind, and they are very light. In fact they contain no *endosperm.* As this is the substance that provides food for seedlings, growing orchids from seed is difficult. The difficulty is increased by the fact that the embryo can grow only when 'infected' with a kind of fungus.

Most orchids do not need this fungus once they have started to grow. However with a few there is life-long dependence. This is the case with some orchids which have little or no chlorophyll. These plants known as *saprophytes* can be seen growing in the heavy shade of the floor of the rain forest. They are brownish or yellowish in all parts; their leaves are no more than scales. An example is *Auxopus,* discussed on page 29.

We have seen that most orchids are pollinated by insects. Some have aids that attract insects. For instance there are genera with flowers that look like the females of certain species of insects and this attracts the males, see figure 7. Other orchids are scented. However a smell that insects like may not always appeal to us! Thus there are several species of the genus *Bulbophyllum* with a rather unpleasant smell like that of dead animals.

It is believed that the Orchid family, although very large, is quite recent in origin. Certainly in some ways there is evidence of an ongoing evolution process. Thus it is easy to create new hybrids. This is often done in commercial horticulture. Also many existing species seem to be rather unstable. Of course this sometimes makes the classification of orchids very difficult.

Fig. 2 (above left). The division of Nigeria in this book into a northern (N) and a southern (S) part coincides, *grosso modo,* with the division between rain forest, in the south, and savanna in the north.

Fig. 3 (above right). *Angræcum birrimense.*

Fig. 4 (centre). Floral parts of different types of African orchid flowers.

Fig. 5 (below left). Flower of *Eulophia cristata,* with one lateral sepal removed to show the column.

Fig. 6 (below right). The top of a column.

TERRESTRIALS AND EPIPHYTES

If we use the habitat as the basis of classificaton, we find that there are two groups of orchids in tropical Africa. These are the *terrestrials* and the *epiphytes.*

Terrestrial orchids root in the earth, from which they get water and minerals, see figure 8. Usually they rest and wither during the dry season. All the orchids of temperate climates are terrestrial.

The epiphytes grow on trees, see figure 9. Often they are found high up in the crown where they can obtain a lot of light. However some prefer

* A number of orchids are more or less regularly self-pollinated.

4 *Ophrys apifera*

the lower, deeply shaded branches of trees in the tropical rain forest. Epiphytes have roots to hold them where they grow, and also free (or *aerial*) roots. The main use of the latter is to get moisture from the air.

It is important to note that epiphytic orchids are not parasites – the trees on which they grow give them no food, only support. (Strictly speaking, though, the orchids are parasites in relation to the fungus necessary for the germination of the seeds.) The minerals that the epiphytes need are provided in the water that runs down the bark when it rains. Clearly this is a limited source of nourishment, so these plants grow rather slowly.

The roots of epiphytic orchids generally have a thick cover of dead cells. This is called the *velamen.* It absorbs water well by capillary action, so that the plant can make the best use of the available moisture.

The velamen is also a good insulator. When the velamen is dry, it is usually light grey in colour. However it becomes rather transparent when wet. Then the layers below can be seen through it – making the roots look green. This is because the roots have chlorophyll; they are assimilating roots, an unusual feature in a plant.

During the dry season, most epiphytes lose some leaves, but it is rare for them all to go. As there are often long periods without rain, most of these plants have methods of storing water. In some cases, the lower part of the stem is swollen. The swelling is called a *pseudobulb,* see figure 10. In other species, the leaves are thickened and act as reservoirs.

The tropical rain forest is richest in epiphytic orchids. Many are quite particular in their choice of a host. *Ansellia,* the stately Leopard Orchid, for instance, seems to prefer *Borassus* and *Hyphaene* palms. Some trees never support orchids. These include trees with latex (like the rubber tree and the mango*) and other introduced species such as *Eucalyptus.* Trees with moderately rough bark will provide a good surface for orchid roots. So epiphytes are often found on *Parkia,* the Locust-Bean tree, for instance.

THE NAMES OF ORCHIDS

As Nigerian orchids are very little known, they rarely have indigenous names. Instead, the *binomial system* is used. In this system, any plant is denoted by two latin (or latinized Greek, etc.) words, as for instance *Angraecum subulatum.*

The first word is the *generic name,* which is shared by all the plants in the same genus.

In our case above, the name *Angraecum* is derived from the Malay word 'anggrek', which means 'orchid'. The second word, the *specific epithet,* is used only for the one species in question. Here, *subulatum* refers to the leaves of this species, they are very narrow and sharply pointed: *subulatum* means *awl-shaped.*

Other common specific names are:

alba = white	*paludosa* = growing in swamps
alta = tall	*parva* = small

Fig. 7 (above left). *Ophrys apifera,* an orchid from Spain, with insect-like flowers.

Fig. 8 (above right). *Eulophia cristata,* a terrestrial orchid.

Fig. 9 (centre). An epiphytic orchid: *Cyrtorchis arcuata.*

Fig. 10 (below). An orchid with pseudobulbs: *Bulbophyllum* sp.

* In Asia, however, indigenous orchids are frequently found on these trees!

aphylla = without leaves	*parviflora* = with small flowers
flavida = yellow	*paucifolia* = with few leaves
odorata = fragrant	*parvifolia* = with small leaves
maculata = spotted	*pubescens* = with fine hairs

For all the specific names above, the feminine form, ending in -a has been given. This is used when the generic name is a feminine one. If the generic name is masculine or neuter, the endings '-us' and '-um' are used instead.

It may seem surprising that Latin, a dead language, should be used in the naming of plants that are very much alive. Moreover, Latin plant names tend to be long and at times rather difficult to pronounce. But Latin was the *lingua franca* of scientists at the time when the binomial system was invented. Any scientist could communicate with his colleagues in other countries by using this language.

Even today, Latin plant names have many advantages, the main one being that a species always has the same name, wherever it may be growing. This is not the case with indigenous names. An English bluebell is very different from a Scottish one, for instance. But if you call them *Hyacinthoides nonscriptus* and *Campanula rotundifolia,* respectively, there will be no mistake.

In order to make quite sure that everyone understands which species a binomial name refers to, the name of the *author,* i.e. the person who gave the name, is often stated together with the plant name in scientific works, thus: *Angraecum subulatum* Lindl. This binomial has been attributed to the orchid in question by John Lindley, a famous 19th century botanist.

Sometimes, the scientific name of a plant has had to be changed for some reason or other. The rule is then that the original specific epithet is kept, together with a new generic name, and the name of the first authority is given in brackets. For instance, in 1866, Reichenbach the Younger (Rchb.f.) described an orchid which he called *Eulophia saundersiana.* Much later, in 1957, V.S.Summerhayes revised the genus and found that the plant, together with a few closely related species, could better be placed in a separate genus. After Summerhayes' revision, the plant was known as: *Eulophidium saundersianum* (Rchb.f.) Summerh.

Sometimes, a species has been found, described and named by two persons independently, perhaps in two different places. One of these names will eventually be the official one, but it may take a long time before this decision is reached. Therefore, the same species is for some time known under two names. When, at last, one of them has been retained, the other one, the *synonym* (syn.) is sometimes quoted together with the official name, to avoid mistakes.

Occasionally, when the specific name is not known, or when it is convenient to refer to the genus only, a plant can be designated like this: *Angraecum* sp. Here, sp. stands for species. When it is a question of more than one species, one writes spp. Sp. nov. means species nova – a new species.

Within a species, there is sometimes so much variation that the need arises to distinguish between *subspecies, varieties* and *forms.* This is done by adding subsp., var. and f., plus a suitable Latin attribute to the specific name: *Cyrtorchis arcuata* subsp. *variabilis* Summerh., *Ansellia gigantea* var. *nilotica* (Baker) Summerh.

THE USE OF ORCHIDS

On the whole, the Orchid family has little commercial value. The only extensively exploited member is *Vanilla planifolia* and related species. They are cultivated mainly in the Malagasy People's Republic in the Indian Ocean; from the seed capsules comes vanilla. A number of Central American and Asian epiphytic orchids are grown as ornamental plants. Most are artificial hybrids, selected for flower size and colour in particular.

African orchids are much less known in horticultural circles. However there are some that compare well with the commercial species. In the chapter on Nigerian epiphytes, mention will be made of a few orchids that are well suited for amateur cultivation. Surprisingly enough, terrestrials are, on the whole, more difficult to transplant than epiphytes, even when all possible care is taken to find conditions for them that closely resemble their natural habitat.

As we have seen, orchids are quite demanding as regards habitat and climate. Thus they tend to disappear as a result of forest clearing and bush burning. The most important habitat of epiphytic orchids, the primary lowland rain forest, will have disappeared everywhere in the tropics before the end of this century at the present rate of deforestation, according to the experts.

It is really sad that many of these little-known plants are becoming extinct: they form an integral and unique part of the natural heritage of the country, as much as lions and elephants. Like these, some orchids would need protection and reserves . . .

CHAPTER TWO

Nigerian orchids, terrestrials

There are around 20 genera of terrestrial orchids in Nigeria, with more than 250 species. Many of these are found in the savanna, where they follow a vegetation cycle similar to that of the terrestrials in temperate climates: the vegetation period starts in the early rains, when many of these plants flower; then follows a period of rest, when the plants wither, during the dry season. Most of these ground orchids have a thickened rhizome, by means of which they hibernate, and they are often able to survive bush burning.

There are also genera, *Oeceoclades* (p.41) for instance, with well developed pseudobulbs. Many terrestrials have no leaves at the time of flowering; these often develop much later, at the height of the rainy season.

It is then sometimes quite difficult to identify the particular orchid, as the inflorescence has withered and disappeared, for instance in the case of *Nervilia* species (p.27) and many *Eulophia* species (p.29).

SIMPLIFIED KEY TO THE GENERA OF NIGERIAN GROUND ORCHIDS DESCRIBED IN THIS BOOK

Genera within brackets will not be further treated.

A. Lip without a spur
  1. Dorsal sepal with a spur, petals free from dorsal sepal . . . . . [*Disa*]
  2. Dorsal sepal without a spur
    I. Plants leafless, at least when flowering
      a. Plants without chlorophyll
        i. Flowers white or pale brown, less than 1 cm long; petals and sepals united in lower part. *Auxopus,* p.29
        ii. Flowers lilac, 2 cm long, large for size of plant . . . . [*Schwartzkopffia*]
      b. Plants with chlorophyll; flowers appearing before leaves, more than 1 cm long; petals and sepals not united . . . . *Nervilia,* p.27
    II. Plants with leaves at time of flowering
      a. Lip not attached to column; sepals and petals free, leaves numerous; inflorescence with many, purple flowers . . . . *Brachycorythis,* p.23
      b. Lip attached to column; dorsal sepal and petals united; lateral sepals united at base, each with small pouch near inner margin . . . . *Disperis,* p.25

B. Lip with a spur
  1. Lip not attached to column

I. Stigmas on stalks
a. Flowers yellow or orange; stem slender; leaves narrow, grass-like; lip entire . . . . *Platycoryne,* p.25
b. Flowers greenish or white, seldom yellow, often complicated, insect-like. Petals often 2-lobed, lip 3-lobed . . . . *Habenaria,* p.9
c. Flowers bluish to mauve; stigmas on short stalk . . . . [*Cynorchis*]
II. Stigmas without stalks; stem with many leaves . . . . *Brachycorythis,* p.23
2. Lip attached to base of column
I. Flowering stem separate from leaves, often with pseudobulb
a. Leaves thin, plicate, often not developed at time of flowering . . . . *Eulophia,* p.29
b. Leaves thickened or leathery, often folded, present at time of flowering . . . . *Oeceoclades,* p.41
II. Flowering stem with 1-2 flattened, orbicular leaves at base, lacking a pseudobulb; lip often many-lobed . . . . [*Holotrix*]
C. Lip with two spurs . . . . [*Satyrium*]

GENUS HABENARIA

This genus is a very large one (more than 800 species), with representatives nearly all over the world. There are also closely related genera, e.g. *Platanthera,* in temperate regions. Nigeria has more than 50 species.

Nearly all the species are terrestrial and most inhabit savanna and open grassland. The tuber produces an annual flowering shoot. The stem is unbranched, generally leafy, with a terminal inflorescence which can have one or many flowers.

The flowers in this genus are often very complicated, and it can be quite difficult to distinguish the usual typical segments in the perianth, see figure 11.

The sepals are entire (undivided), but the petals, and especially the lip, are generally much divided. The dorsal sepal often forms a helmet-like structure together with the two petals, or with one part of them.

The column is provided with two more or less club-shaped stalks, *stigmatic processes,* each carrying one of the stigmas. Similarly, the lower part of the anther is drawn out to form *anther processes.*

A third type of process is seen in some species: two rudimentary *staminodes.* They are often very small, but can at times have a length of a few millimetres.

In many species, the spur is long and conspicuous. The colour of the flowers is white or pale green, seldom yellow.

Habenaria procera (Sw.) Lindl. [Plate 1.1; Fig. 12]
Unlike most habenarias, this is normally an epiphyte. It is often found on oil palms in rain forest areas. The plant is 50-60 cm high, with leaves throughout the length of the stem. The inflorescence is rather densely 10-

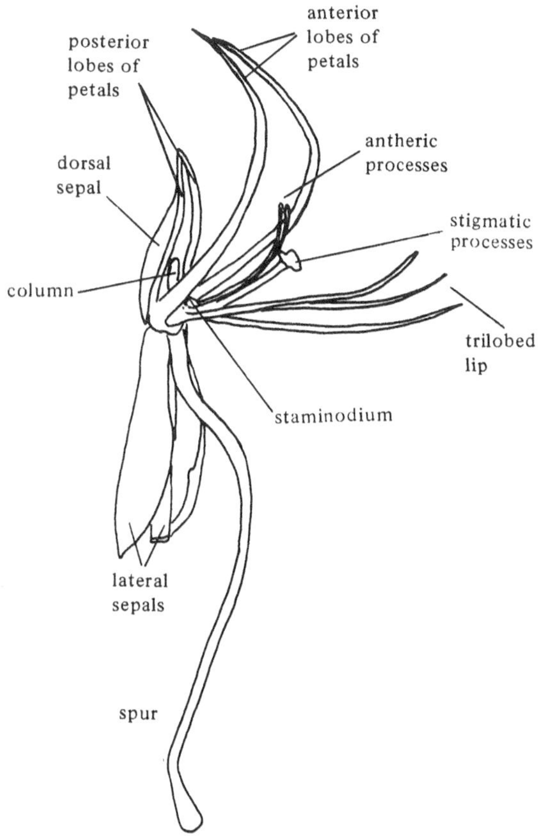

Fig. 11. A *Habenaria*-flower.

Fig. 12. *Habenaria procera.* In forest, on *Elaeis guineensis.* Atta, Imo St. August. 3x.

35-flowered. The hanging spur is conspicuous, 6-11 cm long and swollen in the green, apical part.

Leaves: Lanceolate-acute, 150-300 mm long, 40-70 mm wide. Upper leaves bract-like.

Bracts: Lanceolate or ovate-acute, 17-35 mm long, 6-10 mm wide.

Dorsal sepal: Ovate-acute, 12 mm long, 7 mm wide, much convex, white.

Lateral sepals: Obliquely ovate-acute, 12 mm long, 5-7 mm wide, deflexed, concave, white.

Petals: Oblanceolate-acute, 10-12 mm long, 4-7 mm wide, white.

Lip: Trilobed from common rectangular base; base 5 mm long, 4 mm wide. Middle lobe linear, broader and shorter than side lobes, 18 mm long, 2 mm wide, white. Side lobes linear (subulate), 27 mm long, 1 mm wide, white.

Stigmas: Porrect, club-shaped, green.

Anther processes: Curved upwards, white.

Ovary: 50 mm long.

Distribution: S.

Habenaria gabonensis Rchb.f.

A terrestrial, 15-30 cm high, with a few leaves in the lower part of the stem. The inflorescence seems to be normally 2-flowered, the flowers are a pure white and very similar to the flowers in the preceding species.

Flowering time: October.

Leaves: Lanceolate-acute, 50-60 mm long, 12-15 mm wide, V-shaped in section.

Bracts: Broadly lanceolate-acute, opposite, 1.5 mm long, 6 mm wide.

Dorsal sepal: Ovate-acute, much convex, 10 mm long.

Lateral sepals: Obliquely ovate-acute, 12 mm long, 10 mm wide, much deflexed.

Petals: Broadly oblanceolate-acute, 11 mm long, 5 mm wide, more or less fused with dorsal sepal.

Lip: Trilobed from common rectangular base. Base 5 mm long, 3 mm wide. Middle lobe linear to very slightly spatulate, apiculate, 15 mm long, 2 mm wide. Side lobes subulate, 23 mm long, 1 mm wide.

Stigmas: Porrect, upcurved.

Spur: 50-120 mm long, swollen and flattened in apical half, 2 mm wide, 3.5 mm wide near apex; curved.

Ovary: 50 mm long.

Distribution: S.

Habenaria obovata Summerh.

This ground orchid can be found growing in montane grassland. The stem, up to 45 cm high, is leafy in the lower part and has a dense, many-flowered inflorescence. The flowers are small, green, and faintly scented. Flowers in October.

Leaves: Lanceolate, up to 90 mm long and 15 mm wide, mostly in lower part of stem.

Dorsal sepal: 5-7 mm long.

Petals: Obovate-obtuse, white.

Lip: Trilobed, side lobes longer than middle lobe.

Spur: Much longer than lip, much widened at apex.

Fig. 13 (below left). *Habenaria parva.* Grassland, Kufana, Kaduna St. June. 0.6x.

Fig. 14 (below right). *Habenaria longirostris.* In savanna, Kujama, Kaduna St. June. 2x.

Fig. 15. *Habenaria longirostris.* 1x.

Fig. 17. *Habenaria malacophylla.* In forest. Jema'a, Kaduna St. October.

Fig. 16. *Habenaria malacophylla.* 3.7x.

Stigmas: Porrect, slightly curved, club-shaped.
Distribution: S.

Habenaria parva (Summerh.) Summerh. [Fig. 13]
A small terrestrial, 10-12 cm high, easily overlooked. At the base of the very slender stem are four longer, slightly folded leaves, and in the lower third of the stem is a small single, lanceolate-acute leaf, 1 cm long and 0.2 cm wide.
Leaves (at base of stem): Lanceolate-acute, 15-25 mm long, 5 mm wide.
Bracts: Lanceolate-acute, 5-7 mm long, 1 mm wide, green.
Dorsal sepal: Ovate-acute, 5 mm long, 3 mm wide, green.
Lateral sepals: Broadly and obliquely lanceolate-acute, 6 mm long, 2-3 mm wide, green.
Petals: Linear-acute with wide base, 5 mm long, 0.3 mm wide, fused with dorsal sepal, greenish white.
Lip: Trilobed, 11 mm long, with spreading lobes. The middle lobe is again divided into two rounded side lobes and a much smaller central lobe. Main side lobes oblanceolate. All parts of lip white, with irregular margins.
Spur: Linear, very slender, 15-21 mm long, pale green.
Distribution: N.

Habenaria longirostris Summerh. [Fig. 14 & 15]
A stout, many-flowered (about 20) ground orchid, 30-100 cm high, with a leafy stem. The largest leaves are in the middle of the stem. The flowers are faintly scented.
Leaves (in middle of stem): Ovate-lanceolate-acute, 70-160 mm long, 40-80 mm wide.
Bracts: Lanceolate-acuminate, 15-40 mm long, 2-10 mm wide (flattened).
Dorsal sepal: Lanceolate-acuminate, width reduced at half-length, then widening again; 12-18 mm long, 3 mm wide at base; 1 mm wide at place of constriction, 2 mm after widening, pale green with darker veins.
Lateral sepals: Obliquely lanceolate-acute, 16-18 mm long, 4-5 mm wide, green.
Petals: Bilobed in upper two-thirds. Posterior lobe fused with dorsal sepal in lower half, lanceolate-acute, 13-18 mm long, 1.5 mm wide. Anterior lobe very narrowly lanceolate-acute, slightly shorter than posterior lobe, 0.5 mm wide. All lobes white. Margins with long white hairs.
Lip: Trilobed, 21-28 mm long, lobes linear, spreading, green. Middle lobe slightly longer than side lobes.
Spur: 15-19 mm long, slightly swollen at apex, more or less hidden by bract.
Stigmas: Porrect, club-shaped, green.
Anther processes: Subulate, curved upwards, connivent.
Distribution: N.

Habenaria malacophylla Rchb.f. [Fig. 16 & 17]
A ground orchid, 30-100 cm high, growing in dense forest. The leaves are thin and soft as the specific name indicates, and present mainly in the middle part of the stem. The inflorescence has 20-35 small, green flowers.

Fig. 18. *Habenaria bongensium.* In savanna, Kufana, Kaduna St. July. 2.5x.

Fig. 19. *Habenaria genuflexa.* In swamp, Fara Kwai, Kaduna S August. 1x.

Leaves: Oblanceolate (obovate)-apiculate. 50-200 mm long, 15-60 mm wide, dark green.
Bracts: Lanceolate-acute, 9-20 mm long.
Dorsal sepal: Ovate-acute, convex, 3-6 mm long, 3-5 mm wide, green.
Lateral sepals: Obliquely lanceolate-acute, 4-7 mm long, 2-3 mm wide, green.
Petals: Bilobed, narrowly linear from broader, white base, 3-7 mm long, green.
Lip: Trilobed. Middle lobe linear-obtuse, slightly widened at apex, 5-7 mm long. Side lobes linear-acute, slightly longer than middle lobe.
Spur: 9-18 mm long, slightly thickened in middle.
Distribution: S + (N).

Habenaria bongensium Rchb.f. [Plate 1.4; Fig. 18]
A small (up to 50 cm high), 15-20-flowered terrestrial with leafy stem. The longest leaves are in the lower part of the stem, upper leaves are bract-like.
Leaves: Lanceolate-acute, 35 mm long, 10 mm wide.
Bracts: Lanceolate-acute, 10 mm long, 2 mm wide, green.
Dorsal sepal: Elliptical-obtuse, 4 mm long, 3 mm wide, pale green.
Lateral sepals: Broadly lanceolate-obtuse, bent downwards and/or rolled up lengthwise, 5 mm long, 3 mm wide, pale green.
Petals: Bilobed. Posterior lobe fused with dorsal sepal, obliquely lanceolate-acute, 3.5 mm long, 1.5 mm wide, pale green. Anterior lobe linear-obtuse, 2 mm long, 0.5 mm wide, white.
Lip: Trilobed. Middle lobe longly triangular-obtuse, 3.5 mm long, 1.5 mm wide, white. Side lobes linear-obtuse, 2 mm long, 0.5 mm wide, white.
Spur: 8-13 mm long, very slightly thickened at apex.
Distribution: N.

Habenaria genuflexa Rendle [Plate 1.3; Fig. 19]
A 15-20-flowered terrestrial, 20-100 cm high, with 3-10 leaves, the upper leaves resembling bracts. Flowers small, green; fragrant, especially at night.
Leaves: Linear to lanceolate, acute, 50-250 mm long, 5-20 mm wide, slightly V-shaped in section.
Bracts: Lanceolate-acute, 8-25 mm long, up to 5 mm wide (flattened), sometimes withered at time of flowering.
Dorsal sepal: Obovate-obtuse, convex, 5-7 mm long, 2-3 mm wide, green.
Lateral sepals: Obliquely obovate-acute, bent downwards, 7-10 mm long, 2-6 mm wide, green.
Petals: Bilobed, lobes spreading. Posterior lobe linear-acute, 4-7 mm long, 0.8-1 mm wide, white. Anterior lobe narrowly lanceolate-acute, 8-16 mm long, 1-2 mm wide, green.
Lip: Trilobed. Middle lobe linear-acute, 9-13 mm long, 0.6-0.8 mm wide, green. Side lobes linear-acute, 6-11 mm long, 0.5 mm wide, green.
Spur: Distinctly swollen in apical part, twisted through 180° in the middle, 8-13 mm long, green.
Distribution: N + S.
(Syn. *Habenaria anaphysema* Rchb.f.)

Fig. 20. *Habenaria* sp. Savanna, Anara F.R., Kaduna St. September. 2.2x.

Fig. 21. *Habenaria laurentii.* Savanna, Anara F.R., Kaduna St. July. 1.4x.

Habenaria sp. Segerbäck 1292 [Fig. 20]

A ground orchid with leafy stem, 20-40 cm high. The inflorescence is rather loosely 7-12-flowered. All parts of plant pale green, except column.

Leaves: Narrowly lanceolate-acute, 30-60 mm long, up to 15 mm wide.

Bracts: Ovate-apiculate, 15-20 mm long, up to 10 mm wide.

Dorsal sepal: Ovate-acute, much concave, 3-5 mm long, 2-3 mm wide.

Lateral sepals: Obliquely triangular-acute, bent sharply downward, 5-8 mm long, 3-5 mm wide.

Petals: Bilobed. Posterior lobe fused with dorsal sepal, elliptical-apiculate, 4 mm long, 2 mm wide. Anterior lobe narrowly lanceolate-obtuse, 3 mm long, 0.5 mm wide, erect and slightly curved.

Lip: Trilobed. Middle lobe narrowly lanceolate-subacute, bent downward, 7 mm long, 1 mm wide. Side lobes narrowly lanceolate-acute, spreading, 5 mm long, 1 mm wide.

Spur: Distinctly swollen at apex, 8 mm long.

Stigmas: Porrect, widened and truncate at apex, 4 mm long, white.

Anther processes: Subulate, recurved, 3 mm long, white.

Staminodia: Conspicuous, bilobed; posterior lobe subulate, 3 mm long. Anterior lobe small, ligulate.

Habenaria laurentii De Wild [Fig. 21]

A large-flowered terrestrial, 30-70 cm high, with a stout, leafy stem. It has 3-15 flowers in a rather lax inflorescence. The ovary forms an angle of nearly 90° with the pedicel in specimens seen by author. The flowers are faintly scented.

Leaves: Elliptical-acute, 70-200 mm long, 30-70 mm wide. Upper leaves are bract-like.

Bracts: Lanceolate-acute, 30-80 mm long, leaf-like, often covering ovary and spur.

Dorsal sepal: Elliptical-subacute, very convex, 20-25 mm long, 10 mm wide, green.

Lateral sepals: Very obliquely semi-obovate, apicular, bent downwards and rolled up lengthwise, 25-30 mm long, 10-13 mm wide, green.

Petals: Bilobed. Posterior lobe fused with dorsal sepal, narrowly linear, 20-22 mm long, green. Anterior lobe linear at base, apical part subulate, 40-85 mm long, 2 mm wide at base. Apex white, basal part green. The anterior lobe points upwards like a horn.

Lip: Trilobed. Middle lobe linear, tapering, 25-40 mm long, 1.2 mm wide, white with green apex. Side lobes linear, tapering, 20-38 mm long, white, with slightly upturned apices.

Spur: 55-80 mm long, with slight S-bend. Apex much swollen, forming blunt sac, 3-5 mm wide.

Stigmas: Porrect, 15-17 mm long, apex widened and truncate.

Anther processes: Slender, 20 mm long, curved upwards in front.

Distribution: N + S.

Habenaria holubii Rolfe [Fig. 22]

A stout, leafy, large-flowered terrestrial, 40-80 cm high. It is very similar to *Hab.laurentii,* and should perhaps rather be treated as a subspecies of this. *Habenaria holubii* has a rather lax, 3-15-flowered inflorescence. The pedicel, ovary and dorsal sepal are in a straight line, and the lateral, tightly

Fig. 22. *Habenaria holubii.* Swamp, Kujama, Kaduna St. August. 1.6x.

Fig. 23. *Habenaria walleri.* Savanna, Anara F.R., Kaduna St. July. 0.9x.

rolled-up sepals form a typical 'St Andrew's cross' together with the anterior petal lobes. Fragrant.

Leaves: (Oblong)-lanceolate-acute, 70-150 mm long, 20-45 mm wide, often pressed close to stem except for topmost 10 mm.

Bracts: Leafy, lanceolate-acute, apex bent downward, 20-60 mm long, 17-22 mm wide, covering lower half of spur but not apex; ovary not covered.

Dorsal sepal: Elliptical to lanceolate, acute, very convex, 15-24 mm long, 6-10 mm wide, green.

Lateral sepals: Very obliquely ovate-acute, rolled up lengthwise and bent downward, 18-27 mm long, green.

Petals: Bilobed. Posterior lobe fused with dorsal sepal (upper ⅔), 15-22 mm long, 0.5 mm wide, linear-acute, white. Anterior lobe subulate, upturned like a horn, 20-40 mm long, 3 mm wide near base, green, apex white.

Lip: Trilobed. Middle lobe linear-obtuse, bent downward, 20-37 mm long, 1 mm wide, greenish white. Side lobes lanceolate-acute, straight, 6-20 mm long, 2-3 mm wide, green.

Anther processes: Very slightly upcurved, parallel to, and nearly touching stigma.

Stigmas: Porrect, 13-21 mm long, slightly shorter than anther processes, apex widened and truncate.

Spur: More or less parallel with pedicel and ovary, only slightly curved, 55-80 mm long, much swollen at apex.

Distribution: N.

Habenaria walleri Rchb.f. [Plate 1.5; Fig. 23]

This is an impressive ground orchid, 40-100 cm high, with 3-10 large, attractive flowers. There is not much spreading of the leaves which are more or less closely pressed to the stem. The flowers are fragrant, especially at night, and have a spicy scent, resembling cardamon.

Leaves: Lanceolate-acute, 60-140 mm long, 10-20 mm wide, upper leaves similar to bracts.

Bracts: Broadly lanceolate-acuminate, very thin, 20-110 mm long, 25-35 mm wide.

Dorsal sepal: Ovate-obtuse, sometimes acuminate, very convex, erect, 12-16 mm long. 9-12 mm wide, green.

Lateral sepals: Obliquely semi-orbicular-acuminate, bent downward, 8-20 mm long, 6-9 mm wide, green.

Petals: Bilobed. Posterior lobe fused with dorsal sepal, obliquely subulate, 11-16 mm long, 1-2 mm wide, green. Anterior lobe lanceolate-acute, bent downward, 20-30 mm long, 4-8 mm wide, white.

Lip: Trilobed. Middle lobe oblanceolate-obtuse, 15-32 mm long, 3-7 mm wide, white. Side lobes linear, subacute, 19-45 mm long, 3-7 mm wide, white.

Spur: Hanging, 130-190 mm long, apical half very slightly swollen and flattened, 3 mm wide, tapering at apex. Normally hidden by bract.

Stigmas: Porrect, club-shaped, 8-14 mm long, white.

Distribution: N.

Habenaria occidentalis (Lindl.) Summerh. [Plate 1.2; Fig. 24 & 25]

A rather unusual-looking *Habenaria* with 10-20 yellow flowers which are

Fig. 24. *Habenaria occidentalis*, leaf.

Fig. 25. *Habenaria occidentalis*. In savanna, Anara F.R., Kaduna St. June. 2.5x.

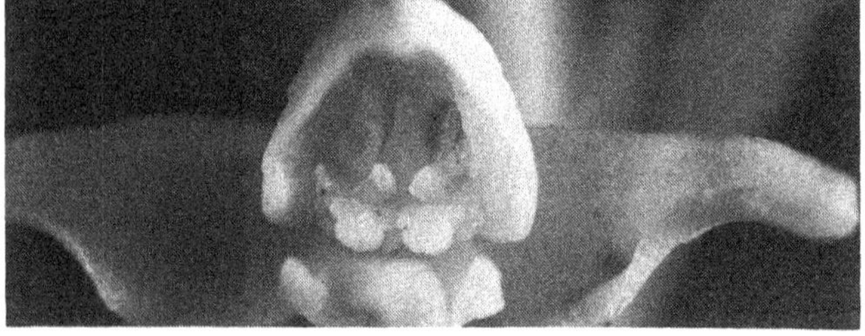

Fig. 26a. *Habenaria lelyi*. 5.7x.

Fig. 26. *Habenaria lelyi*. In savanna, Kufana, Kaduna St. July-August. 1x

not resupinate (the lip points upwards). It is a small ground orchid, 10-30 cm high, and has a slender stem with a single ovate leaf pressed closely to the ground. The whole plant, except for the petals and the column, is densely covered with short hairs.

Leaf: Ovate-acute, cordate, 30 mm long, 27 mm wide, dark green with white reticulate pattern.

Bracts: Lanceolate-acute, 11 mm long, pale green.

Dorsal sepal: Ovate-subacute, 6 mm long, 4 mm wide, pale green.

Lateral sepals: Ovate-acute, spreading, 6 mm long, 4 mm wide, pale green.

Petals: Orbicular-apiculate, with wavy margins, 6 mm long and wide, yellow.

Lip: Bilobed, with thickened keel in middle, ending in point (sometimes very small), 12 mm long, 9 mm wide, yellow, with irregular margins.

Spur: 10-15 mm long, with curved, slightly thickened apex. Faintly and rather unpleasantly scented.

Distribution: N.

Habenaria lelyi Summerh. [Fig. 26 & 26a]

This surprising *Habenaria* often has abnormal flowers. There is sometimes no spur, and the side lobes of the lip can be more or less suppressed. It is a 30-60 cm high, 5-13-flowered terrestrial with a single thick leaf pressed to the ground. All the petals are white and fleshy.

Leaf: Suborbicular-acute, 40-60 mm long.

Bracts: Lanceolate-acute, much shorter than ovary, 20-25 mm long, 6-7 mm wide.

Dorsal sepal: Lanceolate-obtuse, 9-10 mm long, 4-5 mm wide.

Lateral sepals: Obliquely lanceolate-obtuse, concave, spreading, 10 mm long, 5 mm wide.

Petals: Bilobed. Posterior lobe fused with dorsal sepal, lanceolate-acute, 7 mm long, 2 mm wide. Anterior lobe tooth-like, 3 mm long, 1 mm wide.

Lip: Trilobed, curved upwards. Middle lobe lanceolate-obtuse, 11 mm long, 5 mm wide. Side lobes lanceolate-acute, 7 mm long.

Spur (if any): 30-60 mm long.

Distribution: N.

Habenaria keayi Summerh. [Fig. 27]

Small, one-leaved terrestrial; 6-8-flowered. Scape 20 cm long, densely pubescent; hairs 2-3 mm long. One lanceolate-acuminate sheath, 1.5 cm long, pubescent.

Leaf: 75 x 75 mm long, heart-shaped, apiculate, sessile; adpressed to ground; pale green, with darker reticulate pattern. Upper side densely pubescent, lower side glabrous.

Bracts: Lanceolate-acuminate, 10 mm long, densely pubescent.

Dorsal sepal: Ovate-subacute, 7 mm long, 3 mm wide, green; dorsal side densely pubescent; not reflexed.

Lateral sepals: Slightly obliquely lanceolate-acute, 7.5 mm long, 3 mm wide; green; dorsal side densely pubescent; much reflexed.

Petals: Bilobed, glabrous. Posterior lobe narrowly lanceolate-acute, adnate to dorsal sepal; 7 mm long, 1.5 mm wide at base; pale green with white

Fig. 27. *Habenaria keayi.* Gallery forest, Kajuru, Kaduna State. June. 2.5x.

Fig. 28. *Brachycorythis pubescens.* In seasonal swamp, Kujama, Kaduna St. July. 2.5x.

base. Anterior lobe linear, curved upwards; 12 mm long, 0.6 mm wide; pale green.

Lip: Trilobed with white base; glabrous. Mid-lobe linear-obtuse, reflexed; 10 mm long, 1 mm wide, pale green. Side lobes longly linear, 12 mm long, 1 mm wide; much spreading; apices up-curved; pale green.

Spur: With slight S-bend; swollen near apex; 15 mm long; glabrous.

Stigmas: Nearly sessile; apices obtusely and irregularly swollen; green.

Anther processes: Subulate, porrect; pale cream, 2 mm long. Staminodes ligulate, hyaline, 1 mm long.

Ovary: 13-15 mm long; green; densely pubescent.

Distribution: N + S.

GENUS BRACHYCORYTHIS

There are some 30 Asian and African terrestrials in this genus, usually with numerous leaves. Eight species have been recorded for Nigeria, one of them epiphytic. The flowers are sometimes spurred, but more often, as in the two species described below, the spur is replaced by a sack- or boat-shaped structure in the basal part of the lip. The apex of the lip is generally trilobed, with a smaller middle lobe.

Brachycorythis pubescens Harv. [Fig. 28]

A many-flowered ground orchid, 25-80 cm high, with numerous leaves; hairy in all parts except petals and column. Inflorescence rather dense, 2.5 cm wide. The leaves are very slightly longer in the middle part of the stem.

Leaves: Lanceolate-acute, 30-60 mm long, 10-25 mm wide.

Bracts: Longer than flowers at base of inflorescence.

Dorsal sepal: Elliptical-obtuse, 5-7 mm long, 2 mm wide, purple with dark apex.

Lateral sepals: Obliquely lanceolate-acute, 5-8 mm long, 2.5 mm wide, purple with darker apex.

Petals: Lanceolate-acute, touching each other and forming a hood-like structure with the dorsal sepal; 4 mm long, 1.5 mm wide, purple with darker apex.

Lip: Trilobed. Side lobes broadly and obliquely obovate-obtuse, middle lobe smaller, oblong-acute. Front part of lip 5-10 mm long, 5-14 mm wide, purple. Lower part of lip narrow, 1.5 mm long, yellow with purple spots.

Distribution: N.

Brachycorythis ovata ssp. *schweinfurthii* (Rchb.f.) Summerh. [Fig. 29]

A sturdy, many-flowered terrestrial, 40-100 cm high, with numerous leaves along the stem. The inflorescence is about 5 cm wide. The longest leaves are in the middle part of the stem, the lower surface of the leaves having three prominent veins.

Leaves: Oblong-lanceolate-acute, 65-80 mm long, 20-25 mm wide, pale green with purple tips.

Bracts: Oblong-lanceolate-acute, 40 mm long, 10 mm wide, pale green with purple tips. Lower bracts nearly as long as flowers.

Fig. 29. *Brachycorythis ovata* ssp. *schweinfurthii.* In savanna. Kujama, Kaduna St. August. 0.75x.

Fig. 30. *Platycoryne paludosa.* In swamp, Kujama, Kaduna St. July. 3.5x.

Plate 1
Above left: *Habenaria procera* (Sw.) Lindl. 3.2x.
Above centre: *Habenaria occidentalis* (Lindl.) Summerh. 3.5x.
Above right: *Habenaria genuflexa* Rendle 5.1x.
Below left: *Habenaria bongensium* Reichb.f. 5.1x.
Below right: *Habenaria walleri* Reichb.f. 3.2x.

PLATE 1

Dorsal sepal: Ovate-obtuse, 5-8 mm long, 4 mm wide, concave, forming a hood with petals, pale purple.
Lateral sepals: Obliquely ovate-obtuse, 6-10 mm long, 6 mm wide, pale purple with darker middle vein, slightly spreading.
Petals: Obliquely triangular-obtuse, close together, 10 mm long, 6 mm wide, white with pale purple middle vein.
Lip: Trilobed. Side lobes erect in basal part of lip, bent downwards in apical part. Middle lobe much smaller than side lobes, obtuse, with a small notch at the tip. Basal part of lip 4-6 mm long, white. Apical part, when flattened, 15 mm long, 12 mm wide, purple.
Distribution: N.

GENUS PLATYCORYNE

These orchids are only found in Africa. They are slender terrestrials with unbranched stems. In all, there are about 17 species, of which two can be found in northern Nigeria. The leaves are usually narrow. The flowers, in a few-flowered inflorescence, are often a bright yellow or orange.

Platycoryne paludosa (Lindl.) Rolfe [Fig. 30]
A very slender, 1-5-flowered terrestrial, 15-45 cm high, with 3-6 leaves. The two upper leaves are pressed to the stem, and small. There is often a very small leaf at the base of the stem. The flowers are fleshy and bright orange.
Leaves: Lanceolate-acute, 40-70 mm long, 5-9 mm wide (when flattened).
Bracts: Lanceolate-acute, 12-15 mm long, covering about half of ovary.
Dorsal sepal: Ovate-acute, convex, 7-9 mm long, 5 mm wide.
Lateral sepals: Lanceolate-acute, bent downwards, 7-10 mm long, 2-2.5 mm wide.
Petals: Fused with dorsal sepal, slightly obliquely lanceolate-acute, 8 mm long, 2.5 mm wide.
Lip: Linear-obtuse, slightly spatulate, bent downward, 6-7 mm long, 1.2 mm wide.
Spur: 8-10 mm long, apex bent slightly upwards and thickened.
Distribution: N.

GENUS DISPERIS

This is a fairly large-sized genus (about 75 species), growing in the Old World tropics from Africa to New Guinea. Eight species have been recorded for West Africa. Most species are small, few-leaved ground orchids with a preference for a shady habitat. The anatomy of the flowers is very complicated, especially that of the lip, which is attached to the column by its 'claw' and continues above the column in the shape of a lobed or frilled appendage. The dorsal sepal is united with the petals. Each of the conspicuous lateral sepals (together looking rather like a bilobed lip) has a more or less pronounced pouch near the inner margin. The basal parts of the sepals are united along this margin.

Fig. 31. *Disperis johnstonii.* In gallery forest, Kufana, Kaduna St. August. 2.5x.

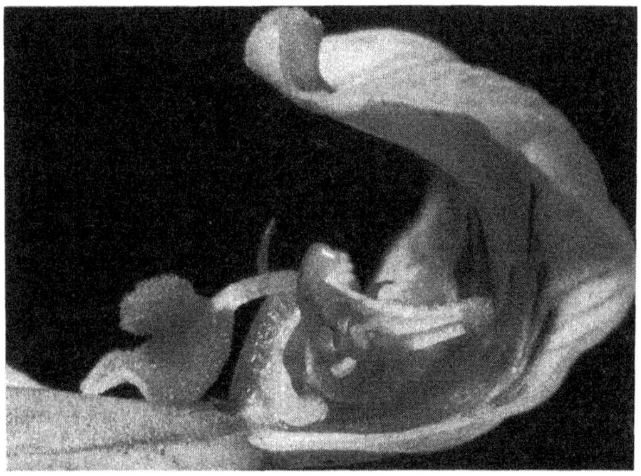

Fig. 32. *Disperis johnstonii.* 9x.

Fig. 33. *Disperis johnstonii,* leaf. 6x.

Fig. 34. *Nervilia umbrosa,* leaf. 0.15x.

Disperis johnstonii Rchb.f. ex Rolfe [Plate 4.4; Fig. 31 & 32 & 33]

A small, 2-leaved terrestrial, 7-13 cm high, with 2-6 flowers.

Leaves: Rounded ovate-acute, cordate, sessile, alternate, 13-25 mm long, 7-15 mm wide, papillose on both sides. Upper surface dark green tinged purple, with white reticulate pattern; lower surface dark purple with marked veins.

Bracts: Ovate-lanceolate-acute, 3-11 mm long, 1-7 mm wide.

Dorsal sepal and petals: Petals fused with dorsal sepal, forming a hood-like structure, 7-10 mm wide, 7-12 mm long. Petals pale purple with darker streak near dorsal sepal. Inner side of petals with irregular yellow or green streaks or spots, bordered with dark purple.

Lateral sepals: Obliquely ovate-subacute, 7-14 mm long, 4-6 mm wide, white with pale purple streak.

Lip: Transversely elliptical-acuminate with marked keel near claw. Appendage of lip bilobed, with divergent lobes. Appendage and blade of lip yellow, claw white. Blade 2 mm long and wide. All parts of column purple or pale purple.

Distribution: N.

GENUS NERVILIA

This genus, comprising about 80 members, has six species recorded for Nigeria, mainly in the north. The single, plicate, kidney-shaped or oval leaf appears some time after the flowering. The short-lived flowers are rather *Eulophia*-like, and greenish, often with purple veins or markings. The lip is trilobed, with small side lobes.

Nervilia umbrosa (Rchb.f.) Schltr. [Plate 4.3; Fig. 34 & 36]

A terrestrial plant, 25-40 cm high, leafless at the time of flowering except for two leaf-like sheaths, 4-5 mm long. The inflorescence is 2-10-flowered.

Leaf: Orbicular-acuminate, cordate, 100-150 mm long, 120-170 mm wide, petiole 200 mm long.

Bracts: Linear-acute, with incurved margins, 20-25 mm long, 2.5 mm wide at base when flattened.

Dorsal sepal: Linear-oblong-subacute, apex curving slightly upwards, 25-40 mm long, 5-6 mm wide, green, tinged purple, with longitudinal purple streaks.

Lateral sepals: Linear-oblong-subacute, 25-40 mm long, 4-5 mm wide, apex curved slightly upwards; green, tinged with purple.

Petals: Linear-oblong-subacute, apex curving slightly upwards, 25-35 mm long, 5-6 mm wide, greenish purple with purple veins.

Lip: Trilobed, lobes triangular-subacute. Middle lobe 13 mm long, apex much recurved. Whole lip 25-40 mm long (with straightened apex), 9-15 mm wide (middle lobe), with two longitudinal purple keels from base of lip to base of middle lobe. Keels continue, flattened, to apex. Middle lobe thickened, green. Whole lip pale green with purple veins.

Ovary: Five-winged, the three wings corresponding with sepals being more pronounced.

Distribution: N + S.

Fig. 35. *Nervilia kotschyi,* leaf. 0.3x.

Fig. 36. *Nervilia umbrosa.* In savanna, Anara F.R., Kaduna St. April-May. 2.5x.

Fig. 37. *Nervilia kotschyi.* In savanna, Anara F.R., Kaduna St. April-May. 2.5x.

Nervilia kotschyi (Rchb.f.) Schltr. [Fig. 35 & 37]

A terrestrial plant, 10-25 cm high, 2-5-flowered, leafless at the time of flowering except for two small, purple sheaths.

Leaf: Ovate-apiculate, cordate, 50-170 mm long, 90-130 mm wide, petiole 10-60 mm long.

Bracts: Very narrowly lanceolate, 10-12 mm long, 1 mm wide at base.

Sepals: Linear-acute, 11-19 mm long, 3 mm wide, olive green.

Petals: Linear-oblong-subacute, 9-17 mm long, 4 mm wide, greenish mauve. All tepals curved slightly backwards.

Lip: Trilobed, 10-19 mm long, 4 mm wide (at base of middle lobe). Side lobes triangular, erect, slightly recurved. Middle lobe 4-6 mm long, triangular, slightly recurved. From base of lip, two longitudinal keels extend to base of middle lobe. Middle vein green, with purple streak at apex of middle lobe. Whole lip cream-coloured or very pale green, with purple veins.

Ovary: Five-winged.

Pedicel: 5 mm long, twisted.

Distribution: N.

GENUS AUXOPUS

There are only two representatives of this small genus in Nigeria. These are small parasites – the whole plant is yellowish brown and incapable of photo-synthesis. It is wholly dependent on the existence of certain fungi. An unusual feature in this genus is the elongation of the pedicels after flowering, so that the seed capsules are borne on long stalks. The habitat of *Auxopus* is the densely shaded soil of the rain forest.

Auxopus kamerunensis Schltr. [Fig. 38 & 39 & 40]

A small terrestrial, about 20 cm high. The leafless stem is very slender, 0.04-0.07 cm wide, and is provided with 2-3 minute scales. When budding, the many-flowered inflorescence forms a compact head. The tepals are united at the base, forming a tube. The bracts are very small, thin and dry.

Dorsal sepal: Lancolate-subacute, 3 mm long, pale brown.

Lateral sepals: United except for apex, forming a notched, ellipsoid disc, 4 mm long, white.

Petals: Lanceolate-subacute, 3 mm long, pale brown.

Lip: Tongue-like, 3 mm long, white.

Capsules: 7-15 mm long.

Distribution: S (+ N).

GENUS EULOPHIA

This is a very large and very 'African' genus: only a few species out of a total of about 200 are to be found in the tropics of America and Asia. In Nigeria, there are some 40 species, many of which are found in the northern part of the country.

Fig. 38. *Auxopus kamerunensis.* In rain forest, Obinze, Imo St. February. 7x.

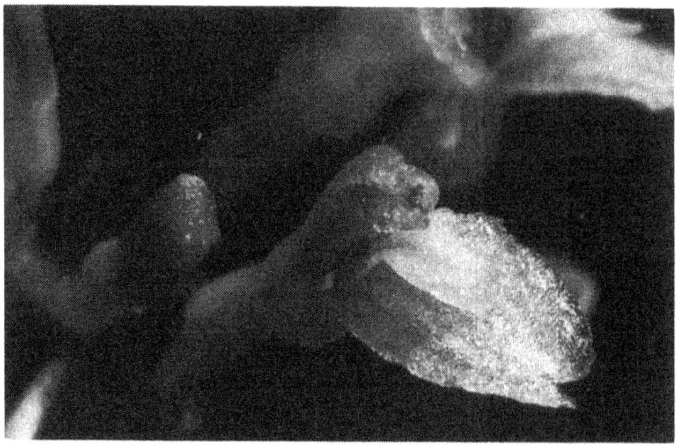

Fig. 39. *Auxopus kamerunensis.* 11x.

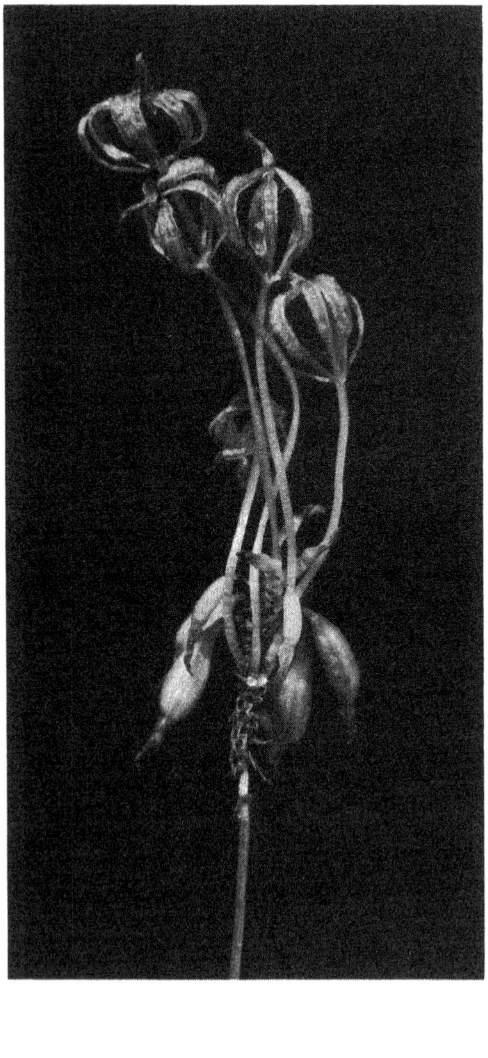

Fig. 40. *Auxopus kamerunensis,* capsules. 2x.

The Eulophias are often quite impressive plants with tall flowering stalks and large, attractive flowers. Many of them have pseudobulbs or tuberous underground stems. The leaves are generally plicate (folded lengthwise like a fan) and often appear after the flowering.

Eulophia quartiniana A.Rich. [Plate 2.3; Fig. 41]

A terrestrial, up to 90 cm high, with a 7-15-flowered inflorescence. The flowering stem has a few, broadly lanceolate-acute sheaths, 3.5-4 cm long, clasping the stem. The perianth segments are shown in Fig. 4b.

Leaves: Ovate-acute, basal, at time of flowering 100 mm long, 40 mm wide; later 250 mm long, 120 mm wide. Plicate.

Bracts: Lanceolate-acute, 10-20 mm long, 2-4 mm wide.

Dorsal sepal: Lanceolate-acute, with margins curving backwards, 20-35 mm long, 5-6 mm wide, pale green to dark mauve.

Lateral sepals: Lanceolate-acute, with margins curving backwards, slightly oblique, 25-40 mm long, 7-8 mm wide, pale green to dark mauve.

Petals: Broadly lanceolate-acute, with margins curving backwards, 23-30 mm long, 8-10 mm wide, pale green to dark mauve.

Lip: Trilobed. Side lobes small, white, folded into contact with column. Middle lobe suborbicular, rounded, often emarginate, 25-45 mm long, 30-45 mm wide, pink with darker fleck near column.

Spur: Slender, 22-40 mm long (from base of ovary), base pale green, anterior part dark mauve.

Column: White, green or purple; operculum yellow with dark purple anterior margin.

Ovary and base of tepals, very dark purple.

Distribution: N (+ S).

Eulophia guineensis Lindl.

This is very closely related to the preceding species, *E.quartiniana.* The main differences are as follows:

E.guineensis	*E.quartiniana*
Fully developed leaves at time of flowering.	Leaves develop after inflorescence.
Flowers at the end of the rainy season.	Flowers in the beginning of the rainy season.
Apex of lip acute.	Lip rounded or emarginate.
Usual habitat: forest.	Usual habitat: savanna.

Eulophia gracilis Lindl. [Fig. 42]

Usually a terrestrial species, height 40-90 cm, many-flowered. Leaves generally developed when flowering. Pseudobulbs conical, 5.5-8.0 cm high, 2-3 cm wide, with one node. Flowering stalk with a few brown scales, the lower ones clasping the stem.

Leaves: Lanceolate-acute, slightly plicate, long and narrow, rather thin.

Bracts: Lanceolate-apiculate, 11-15 mm long, 2-3 mm wide, pale green.

Dorsal sepal: Ovate-acute, 10-12 mm long, 6 mm wide.

Lateral sepals: Ovate-acute, 12-13 mm long, 6-7 mm wide.

Fig. 42. *Eulophia gracilis.* In forest, Avu, Imo St. February. 2x.

Fig. 41. *Eulophia quartiniana.* In savanna, Kujama, Kaduna St. April-May. 0.9x.

Fig. 43. *Eulophia adenoglossa.* In savanna, Anara F.R., Kaduna St. May. 1x.

Petals: Ovate-acute, 8 mm long, 5 mm wide. Sepals and petals (except lip) green with brown streaks or flecks, margins and base green.
Lip: Truncate, margin with long hairs; yellowish white.
Spur: 8 mm long, swollen at apex and with a wide opening; green, with some brown flecks at apex.
Distribution: N + S.

Eulophia adenoglossa (Lindl.) Rchb.f. [Fig. 43]
A many-flowered terrestrial, 50-75 cm high, with partly developed, grass-like leaves at the time of flowering. The scape has 3-4 cm long, scarious, acute sheaths, up to 10 cm long. The flowers are hanging, and the bracts long and conspicuous.
Leaves: Very narrowly linear-acute, stiff and fleshy, V-shaped or nearly triangular in section, 100-200 mm long at the time of flowering.
Bracts: Narrowly lanceolate-acuminate, 25-45 mm long, 3 mm wide, pale green, apex often withered.
Dorsal sepal: Oblong-acute, 20-22 mm long, 5-6 mm wide, yellowish green, slightly recurved. Middle vein marked, slightly darker than sepal.
Lateral sepals: Irregularly oblong-acute, 22-25 mm long, 6 mm wide, slightly spreading, yellowish green.
Petals: Oblong-acute, or apiculate, 16-18 mm long, 5 mm wide, yellowish green.
Lip: Trilobed, about 20 mm long. Side lobes small, rounded, erect, very pale green with purple streaks. Hypochilium pale green, with two low keels, spreading towards blade. Middle lobe semi-elliptical, emarginate, with wavy margin and many irregularly broken-up keels. Margin with continuous band of brownish purple markings.
Spur: Very short, blunt, 2-3 mm long, yellowish green.
Column: Pale green; operculum cream and purple.
Distribution: N.

Eulophia lindiana Kraenzl. [Fig. 44]
A small ground orchid, 13-20 cm high, 7-11-flowered. Leaves developed at time of flowering. Base of stem slightly swollen, forming a narrow, conical pseudobulb.
Leaves: 2-6, narrowly linear-acute, 30-45 mm long, 3-5 mm wide.
Bracts: Tapering from a wide base, acute, 6 mm long, 1 mm wide.
Sepals: Linear-acute, 8-15 mm long, 2 mm wide, brownish purple.
Petals: Oblong-lanceolate-acute, 7 mm long, 3 mm wide, pale green.
Lip: Trilobed. Side lobes erect or touching, small, rounded, white with purple margins. Middle lobe obovate, with very wavy margin, white. Length of lip 20 mm. Basal part of lip with 3-4 prominent pale green keels.
Spur: 5-6 mm long, 1-2 mm wide at slightly swollen apex.
Column: Pale green; operculum pale green or purple.
Distribution: N.

Eulophia warneckeana Kraenzl. syn. *E.microdactyla* Kraenzl. [Fig. 45 & 46]
Small, slender terrestrial, 15-25 cm high, leafless at time of flowering.

Fig. 45. *Eulophia warneckeana.* 5x.

Fig. 46. *Eulophia warneckeana.* In swamp, Kura River, Kaduna St. April. 2.5x.

Stalk 0.1-0.15 cm wide with three clasping, lanceolate-acute sheaths, 3-3.5 cm long. Inflorescence rather laxly many-flowered.

Leaves: Very narrowly linear-acute, plicate, up to 300 mm long, 3 mm wide.

Bracts: Linear-acuminate, 4-6 mm long.

Tepals: Subovate-apiculate (or acute), slightly recurved, 5-6 mm long, 2-3 mm wide, yellow.

Lip: Trilobed, 5 mm long from onset of ovary. Middle lobe pale yellow, elliptical, with lacerate margins and yellow papillae. Side lobes small, indistinctly triangular, slightly upcurved, white. Base of lip white with pale mauve veins and a few mauve papillae.

Spur: Tubular, 3-4 mm long, yellow.

Column: Pale purple; operculum white.

Distribution: N + S.

Eulophia leonensis Rolfe (savanna var.) [Fig. 47]

A terrestrial with partly developed leaves on sterile shoot at time of flowering. Scape up to 50 cm high, many-flowered.

Leaves (at time of flowering): Elliptical-acute, plicate, 50-110 mm long, 30-50 mm wide. Mature leaves up to 300 mm long.

Bracts: Linear-acuminate, 5-10 mm long.

Sepals: Oblanceolate-apiculate, 7-8 mm long, 2.5-3.5 mm wide, purple, or pale green, tinged purple.

Petals: Broadly ovate to suborbicular, obtuse, 5-6 mm long, 5-6 mm wide, cream-coloured (sometimes tinged purple), with green or mauve veins and purple margins.

Lip: Trilobed. Side lobes rounded, erect, base touching column, 1.5 mm long, purple. Middle lobe 9 mm long, 2 mm wide, obtuse, convex, purple, with two white, verrucose keels joining a third, shorter keel at apex of lip.

Spur: Straight, tapering, sometimes very indistinctly trilobed near apex, 5-6 mm long, pale yellow, green, or mauve.

Column: Pale green, sometimes tinged purple. Operculum brown or purple.

Distribution: N.

Eulophia leonensis Rolfe (forest var.) [Fig. 48 & 49]

Terrestrial, leafless at time of flowering (small tuft of acuminate leaves usually emerging at base of stem). Scape up to 45 cm high; lower part with apiculate sheaths, 3-4 cm long. Many-flowered.

Bracts: Subulate, up to 10 mm long.

Sepals: Narrowly obovate-obtuse, 7 mm long, 2 mm wide (laterals 3 mm), green with minute purple spots.

Petals: Broadly elliptical-obtuse, 6 mm long and wide, green with purple veins.

Lip: Distinctly trilobed; middle lobe ligulate-obtuse, 4 mm long, much convex, margin irregular, upturned, with two indistinct, verrucose keels; white with purple base. Side lobes erect, rounded, purple.

Spur: Curved slightly forward, 6 mm long, purple.

Distribution: N.

Fig. 44 (above left). *Eulophia lindiana.* In savanna, Anara F.R., Kaduna St. April-May. 2.5x.

Fig. 47 (below left). *Eulophia leonensis.* In savanna, Anara F.R., Kaduna St. April. 4.5x.

Fig. 48. *Eulophia stenoplectra.* 3.5x.

Fig. 49. *Eulophia stenoplectra.* In forest, Jema'a, Kaduna St. March.

Fig. 50. *Eulophia flavopurpurea.* In seasonal swamp, Buruku, Kaduna St. April-May. 1.75x.

Eulophia flavopurpurea (Rchb.f.) Rolfe syn. *E.millsoni* (Rolfe) Summerh. [Fig. 50]

Terrestrial, 60-75 cm high, 6-12-flowered. Leafless at time of flowering, scape with 4-5 lanceolate-acute sheaths.

Leaves: Linear-acute, stiff and erect, plicate, up to 600 mm long.

Bracts: Lanceolate-acute, 15-22 mm long, 6 mm wide at base.

Dorsal sepal: Oblong-acute, recurved, 22-26 mm long, 5-7 mm wide, pale green.

Lateral sepals: Oblong-acute, slightly falcate, recurved, 27-32 mm long, 7 mm wide, pale green.

Petals: Oblong-obtuse, 22-25 mm long, 8-10 mm wide, pale green, more or less connivent.

Lip: Indistinctly trilobed, 25-30 mm long. Side lobes small, rounded, erect, very pale green. Middle lobe 20 mm long, sharply deflexed, suborbicular, with crenulate, recurved margins, very pale green or pale yellow, with up to 11 pale mauve to purple verrucose keels.

Spur: Short, blunt, 5 mm long from onset of ovary, pale green with darker apex.

Column: Pale green; operculum pale yellow.

Distribution: N + S.

Eulophia cristata (Sw.) Steud. [Plate 2.4; Fig. 51]

A many-flowered (up to 40) terrestrial, leafless at the time of flowering. Scape 70-120 cm high. The flowers are faintly scented. This orchid is quite common in northern Nigeria, where it flowers at the end of the dry season.

Leaves: Narrowly lanceolate-acute, plicate, 20-60 mm wide.

Bracts: Lanceolate-acute, 7-11 mm long, 2-3 mm wide.

Tepals: All similar: oblong-subacute, 16-23 mm long, 7-9 mm wide, slightly recurved, very pale purple, occasionally with darker tips. Paired sepals slightly falcate.

Lip: Trilobed, 13-22 mm long (flattened), 8 mm wide. Side lobes orbicular, 7 mm, green. Middle lobe ovate, purple, margin much wrinkled, with two semi-circular keels, 2 mm high, 1 mm wide at throat, continuing as 4-6 thin and much wrinkled ridges to within 3 mm of sharply upturned apex.

Spur: 5 mm long, tapering towards apex, straight or very slightly upturned.

Column: Green, sometimes tinged purple. Operculum yellow.

Distribution: N + S.

Eulophia angolensis (Rchb.f.) Summerh. syn. *E.lindleyana* (Rchb.f.) Schltr. [Plate 2.5; Fig. 52]

A tall terrestrial, up to 180 cm high, leafy at the time of flowering. Inflorescence branched, many-flowered. The scape has a few lanceolate-acute sheaths, 5.5-7.5 cm long, more or less closely pressed to the stem.

Leaves: Narrowly lanceolate-acute, 500-700 mm long, 20-45 mm wide, slightly plicate, leathery, lower half folded together lengthwise.

Bracts: Lanceolate-acute from narrow base, 9-19 mm long, 2-6 mm wide.

Sepals: Linear-spatulate, much concave, 20-22 mm long, 4-7 mm wide, spreading, yellow or brown.

Petals: Ovate-obtuse, 20-30 mm long, 12-15 mm wide, touching, slightly convex at apex, yellow.

Lip: Indistinctly trilobed, 25 mm long from base of ovary when not flattened. Side lobes erect, rounded, touching column. Middle lobe long, with much deflexed margins and three triangular keels at base, yellow. Base of lip, including side lobes, pale yellow with purple stripes.

Spur: Very short and blunt.

Column: 15 mm long, 5 mm wide, white; operculum white with dark purple spot near rostellum.

Distribution: N + S.

Eulophia cucullata (Sw.) Steud. [Fig. 53]

A very handsome, large-flowered ground orchid, not uncommon in savanna areas. The plant is leafless at the time of flowering. The scape is 50-75 cm high and 3-10-flowered.

Leaves: Very narrowly lanceolate or linear, 250-300 mm long, up to 25 mm wide, plicate.

Bracts: Lanceolate-acuminate, scarious, 12 mm long, 4 mm wide.

Sepals: Lanceolate-acute, 20-22 mm long, 8 mm wide, much reflexed, olive green to mauve.

Petals: Broadly elliptical-obtuse, 22 mm long, 19 mm wide, pink.

Lip: Trilobed, 23-25 mm long, 25 mm wide across side lobes. Side lobes rounded, white to pale pink. Middle lobe bilobed, with two erect, quadrangular calli at base of spur, and one longitudinal ridge in the centre. Whole lip pink, region near calli green.

Spur: Blunt, very short, 5 mm long. Interior of spur yellow, with purple streaks.

Operculum: Pale yellow.

Distribution: N + S.

Eulophia dilecta (Rchb.f.) Schltr.

is a very similar species with larger flowers and broader leaves which are present at the time of flowering.

Eulophia buettneri (Kraenzl.) Summerh.

also resembles *E.cucullata* but has smaller flowers and no calli on the lip. There are no leaves at the time of flowering.

Fig. 51 (above left). *Eulophia cristata.* In savanna. Riga Cikun, Kaduna St. March-April. 1x.

Fig. 52 (below left). *Eulophia angolensis.* In swamp, Kujama, Kaduna St. August. 1.35x.

Fig. 53 (above right). *Eulophia cucullata.* In savanna, Riga Cikun, Kaduna St. May. 0.7x.

Fig. 54 (below right). *Eulophia horsfallii.* Swamp, Owerri, Imo St. May, September. 0.8x.

Eulophia horsfallii (Batem.) Summerh. [Plate 5.3; Fig. 54]

A very tall ground orchid, growing in swamps. It can be up to 250 cm high, and has many large, attractive flowers. The long leaves are present at the time of flowering.

Leaves: Up to 1 800 mm long, and 150 mm wide, plicate.

Sepals: Oblanceolate-apiculate, 17-27 mm long, purple or yellowish green, spreading.

Petals: Broadly ovate-obtuse, with sideways overlapping margins, so that the two petals form a horizontal plane above the column; white to pink.

Lip: Trilobed, 20-40 mm long, 16-40 mm wide. Side lobes broad, rounded, erect, green. Middle lobe rectangular, slightly rounded with wavy margins, purple. Base of lip with two large and one small, semi-orbicular,

Fig. 55. *Oeceoclades saundersiana.* In forest, Obinze, Imo St. February. 1.25x.

Fig. 56. *Oeceoclades saundersiana.* 3x.

Plate 2
Above left: *Polystachya golungensis* Reichb.f. 5.1x.
Above centre: *Podangis dactyloceras* (Reichb.f.) Schltr. 3.7x.
Above right: *Eulophia quartiniana* A.Rich. 2.2x.
Below left: *Eulophia cristata* (Sw.) Steud. 3x.
Below right: *Eulophia angolensis* (Reichb.f.) Summerh. 2x.

PLATE 2

pale purple keels, continued as three equally high, pale yellow irregular lamellae up to the base of the middle lobe.
Distribution: N + S.

GENUS OECEOCLADES

This genus of Eulophia-like terrestrials (some 30 species), has a wide distribution, occurring in Africa as well as in the tropical parts of Asia and America. Two species only are recorded for Nigeria. All African species are provided with well-developed pseudobulbs. The leaves are fleshy, they generally have a long stalk and are not plicate (folded lengthwise) as in most Eulophia species. The lip has a large middle lobe which is itself divided into two large, rounded lobes.

Oeceoclades saundersiana (Lindl.) Garay & Taylor syn. *Eulophidium saundersianum* (Rchb.f.) Summerh. [Plate 4.1; Fig. 55 & 56]
A forest terrestrial with narrowly ovate-conical, prismatic pseudobulbs, 5-20 cm high, 1.2-2.2 cm wide, 1-2-(3)-leaved. The many-flowered scape can be up to 60 cm high.
Leaves: Narrowly ovate-apiculate, blade 100-250 mm long, 40-80 mm wide, dark green, leathery, with stalk about as long as blade.
Sepals: Oblanceolate-obtuse, 15 mm long, 4 mm wide, pale green with brownish streaks.
Petals: Ovate-subacute, 12 mm long, 6 mm wide, pale green with brownish streaks.
Lip: Bilobed. Middle lobe divided into two frontal, broadly spatulate lobes, each 10 mm long, 7 mm wide, very pale green or cream, with irregular purple markings at base of frontal lobes. Side lobes rounded, 6 mm long, 8 mm wide, very pale green or cream, with purplish-brown streaks.
Spur: Short, blunt, 5 mm long.
Distribution: S.

Oeceoclades maculata Lindl. syn. *Eulophidium maculatum* (Lindl.) Pfitz.
A widely distributed terrestrial, growing in forests. Pseudobulbs narrowly conical or cylindrical, up to 3 cm high, always 1-leaved. Flowering shoot up to 35 cm high, 5-15-flowered.
Leaves: Narrowly lanceolate-acute, 100-300 mm long, 20-55 mm wide, stalk much shorter than blade; broadly V-shaped in section, thick and leathery, light green with irregular transverse bands of darker green.
Tepals: Elliptic-obtuse, concave, about 10 mm long, petals broader than sepals, greenish white to mauve.
Lip: Bilobed, with two tooth-like projections at the base. Greenish white with red or pink markings. Middle lobe again divided into two frontal lobes. Side lobes erect, rounded, with red veins.
Spur: Pale brown, inflated at apex, curving forward.
Distribution: N + S.

CHAPTER THREE

Nigerian orchids, epiphytes

In Nigeria, there are some 30 genera of epiphytic orchids, the species numbering in all about 200. They prefer the humid rain forest to the savanna, so southern Nigeria is much richer in epiphytes than the northern part of the country.

There are two main types of epiphytes, those with *sympodial* and *monopodial* growth, respectively.

Epiphytes with sympodial growth have a stem with yearly lateral shoots. The stem is often thickened to form a pseudobulb serving as a water and nutrient reservoir. The inflorescence appears either between the leaves at the top of the mature pseudobulb (*terminal inflorescence*) or arises from the rhizome near the base of the pseudobulb (*axillary inflorescence*).

The leaves are thin or fleshy; the roots, although sometimes abundant, are generally rather thin. Each new shoot only flowers once, but sometimes the leaves on it persist during a couple of vegetative periods. Genera: *Liparis, Ansellia, Polystachya, Ancistrochilus, Bulbophyllum, Graphorkis* and others.

In epiphytes with monopodial growth there is a single, slowly growing stem without pseudobulbs. Instead, the leaves (whenever present) are nearly always thickened, and have a water-storing function. The roots are usually quite thick and tough, and the inflorescence is always axillary. The leaves are not shed once a year, as in many sympodial orchids, but persist for several seasons. Genera: all the so-called *Angraecoids – Calyptrochilum, Angraecum, Microcoelia, Diaphananthe, Chamaeangis, Rangaëris, Cyrtorchis,* etc.

SIMPLE FIELD KEY TO THE GENERA OF NIGERIAN EPIPHYTES IN THIS BOOK

Genera within brackets will not be further treated.

A. Plants generally with pseudobulbs, growth sympodial
  1. Plants leafless at time of flowering
    I. Pseudobulbs large, 5-9 cm high, 2-5 cm wide . . *Graphorkis,* p.69
    II. Pseudobulbs small, up to 0.5 cm wide. *Polystachya saccata,* p.49
  2. Plants leafy when flowering
    I. Flowering stalk arising from top of pseudobulb
      a. Leaves generally thin, plicate; column long
      and curved . . . . *Liparis,* p.45
      b. Leaves and column not as above
        i. Lateral sepals forming helmet-like structure;

lip pointing upward, see Fig. 4d . . . . *Polystachya,* p.47
ii. All tepals similar, yellow with brownish spots; pseudobulbs very long . . . . *Ansellia,* p.47

II. Flowering stalk arising from base of pseudobulb
a. Leaves thin, plicate . . . . *Ancistrochilus,* p.57
b. Leaves generally thickened, not plicate
i. Flowers resupinate . . . . *Bulbophyllum,* p.59
ii. Flowers not resupinate. Polystachya-like; petals very small, as in Fig. 4d . . . . (*Genyorchis*)

B. Plants without pseudobulbs, growth monopodial
1. Leafless plants with assimilating roots . . . . *Microcoelia,* p.77
2. Plants with linear, strap-shaped or narrowly elliptic, often thickened leaves, well-spaced along woody stem
I. Lip entire (not distinctly lobed)
a. Flowering stalk with single or well-spaced flowers
i. Lip much broader and different from tepals
1. Lip very concave, often pointed, base enveloping column, gradually passing into conical spur . . . . *Angraecum,* p.73
2. Lip concave, short, triangular, Apex of spur often swollen to globular sac. Flowers in pairs or groups of three. . *Chamaeangis,* p.81
3. Lip flat, with irregular edge, often with a point. Flowers delicate, transparent . *Diaphananthe,* p.79
ii. Lip similar to other perianth members, usually slightly broader
1. Lip slightly broader than tepals, curved backwards. Bracts large, brown, triangular. Spur long, tapering (Fig. 4f). . . . . *Cyrtorchis,* p.87
2. Lip slightly larger than tepals, more or less triangular, sometimes indistinctly trilobed. Spur long, tubular . . . . *Rangaëris,* p.85
b. Flowering stalk very short, at base of plant, with flowers in dense, generally head-like inflorescences . . . . *Ancistrorhynchus,* p.91

II. Lip trilobed
a. Side lobes rounded
i. Middle lobe long, acuminate. Spur about 20 cm long, often spirally twisted . . . . *Plectrelminthus,* p.83
ii. Middle lobe rectangular, spur short, constricted in the middle and abruptly bent downwards near swollen apex. . . . . *Calyptrochilum,* p.71
b. Side lobes often tooth-like, lip with auricles on both sides of mouth of spur . . . . *Tridactyle,* p.95

3. Plants with broadly elliptic, ovate or obovate, dark green, tough but not thickened leaves, closely placed on short stem
I. Spur as long as lip; lip much broader than tepals, gradually passing into funnel-shaped spur with

Fig. 57. *Liparis epiphytica.* Mkpụrụ F. Naze, Imo St. June. 2x.

Fig. 58. *Liparis nervosa.* In forest, Avu, Imo St. August. 3x.

thickened apex . . . . *Eurychone,* p.89
II. Spur longer than lip
a. Lip very similar to tepals. Many-flowered inflorescence with white flowers. Spur with a narrow mouth . . . . *Aërangis,* p.83
b. Lip shorter than tepals; flowers small, green. . . . . *Diaphananthe curvata,* p.79

4. Plants with needle-like leaves
I. Lip not distinctly lobed; flowers white . . . . *Angraecum subulatum,* p.77
II. Lip distinctly trilobed; flowers yellow . . . . *Tridactyle tridentata,* p.95

5. Plants with fleshy, opposite, triangular leaves, closely placed on branched stem. Lip deeply concave . . . *Angraecum,* p.73

6. Plants with fleshy, fanwise arranged leaves; stem short
I. Inflorescence shorter than leaves; spur longer than lip . . . . *Podangis dactyloceras,* p.69
II. Inflorescence longer than leaves
a. Spur shorter than lip. Leaves up to 8 cm long, usually shorter. Inflorescence very dense, with many small flowers . . . . (*Bolusiella*)
b. Spur longer than lip. Leaves 6-18 cm long. Lip hastate *Rangaëris rhipsalisocia,* p.87

GENUS LIPARIS

This genus has representatives practically all over the world, in all, it numbers some 250 species. In temperate regions, Liparis is of course a genus of terrestrial species, but there are a couple of ground orchids also among the seven species recorded for Nigeria. The two species described in this book are mainly epiphytes, however.

Among the characteristic features of the genus are the small pseudobulbs, a long, curved column and, in some species, thin, plicate leaves.

Liparis epiphytica Schltr. [Fig. 57]

A small rain forest epiphyte, 9-11 cm high.

Leaves: Oblong-emarginate, 20-110 mm long, 4-15 mm wide, thickened, stalks 10 mm long.

Bracts: Lanceolate-acute, 7-8 mm long, 3-4 mm wide.

Dorsal sepal: Narrowly lanceolate-obtuse, 5-7 mm long, 2 mm wide at base, margins recurved, pale green.

Lateral sepals: Ovate-obtuse, 5 mm long, 3 mm wide, pale green.

Petals: Narrowly linear, 5-7 mm long, 0.3 mm wide, pale green.

Lip: Elliptical-obcordate, 3-5 mm long, 3-4 mm wide, curled back- and downward. Two distinct, tooth-like processes at base. Pale green.

Column: Slender, bent downward at apex, 4 mm long, 1.3 mm wide, pale green.

Distribution: S.

Fig. 59 (left). *Ansellia gigantea* var. *nilotica*. In savanna, Bokkos, Plateau St. March.
Fig. 60 (below). *Ansellia gigantea* var. *nilotica*, on *Borassus*-palm at Bokkos, Plateau St.

Liparis nervosa (Thun.) Lindl. syn. *L.guineensis* Lindl. [Fig. 58]
This species is said to be predominantly terrestrial, but the author has only found it growing as a epiphyte. Epiphytic specimens seen by the author have been as much as 30 cm high, while the terrestrials are reported to be up to 40 cm. The stalk with five sharp wings is level with, or slightly shorter than, the top of the leaves, and many-flowered. The base of the stem is swollen to form a conical-ovoid pseudobulb, up to 4 cm high and 2 cm wide.

Leaves: Obovate-elliptical-acute, plicate, 50-270 mm long, 20-90 mm wide, stalks 10-60 mm long.
Bracts: Lanceolate-acute, 5-15 mm long, 2-3 mm wide, green.
Dorsal sepal: Ovate-apiculate (or acute, obtuse), 5-6 mm long, 1-2.5 mm wide, with recurved margins, green.
Lateral sepals: Obliquely ovate-acute, much reflexed or coiled, 4-6 mm long, 2.5-3.0 mm wide, green.
Petals: Narrowly linear-obtuse, margins recurved, 4-5 mm long, 0.5-1.0 mm wide, green.
Lip: Bilobed, obcordate, curved downward, with two tooth-like calli at base, 4 mm long, 3 mm wide, green, purple at apex.
Column: 3-4 mm long, 1.2-1.5 mm wide, curved downward, pale green. Operculum green.
Distribution: S.

GENUS ANSELLIA

This African genus is one of the very few to have an English name: these orchids are known as 'Leopard Orchids' because of the yellow perianth-segments with dark spots.

Ansellias have very long, narrowly cylindrical pseudobulbs, 10-60 cm high, rising from a dense mat of thin, vertical roots which, at times, can form a clump of respectable dimensions. The long leaves, from the top of the pseudobulb, are leathery. Among them arises the branched flowering stalk, which is many-flowered. Petals and sepals are similar, more or less elliptical; the lip is much shorter than the other segments. Ansellias are easily cultivated and very attractive.

Ansellia gigantea var. *nilotica* (Baker) Summerh.[Plate 5.4; Fig.59 & 60]
Leaves: Narrowly lanceolate, sharply pointed, 150-400 mm long, 20-60 mm wide.
Tepals: Narrowly elliptical-(sub)acute, up to 30 mm long, 10 mm wide.
Lip: Trilobed, shorter than petals; middle lobe slender, pointed, slightly reflexed, with three keels. Length of lip 25 mm, middle lobe 7 mm, yellow. Side lobes erect, rounded, yellowish green or yellow, with red or brownish-red lines.
Distribution: N + S.

GENUS POLYSTACHYA

Pseudobulbs are quite common in this large genus (more than 300 species) of epiphytes, of which only very few representatives are to be found out-

Fig. 61. *Polystachya laxiflora.* On *Pentaclethra* sp., Elele, Imo St. April-July. 2.3x.

Fig. 62. *Polystachya saccata.* On *Nauclea* sp. in forest, Jema'a, Kaduna St. March. 3x.

Fig. 63. *Polystachya coriscensis.* In rain forest, Ikom, Cross River St. October. 1x.

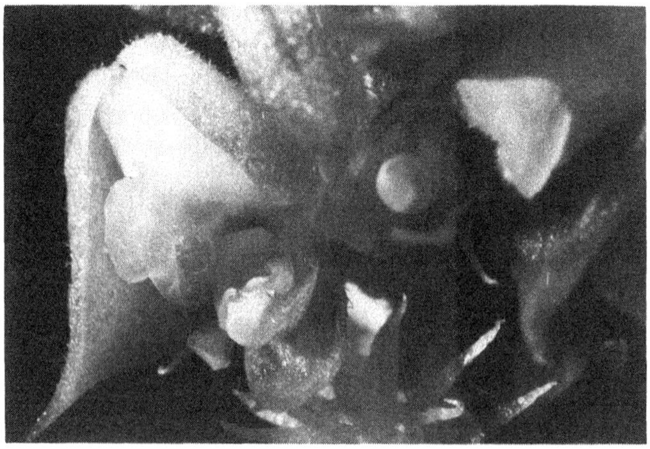

Fig. 64. *Polystachya coriscensis.* 10x.

side tropical Africa. The flowers are very characteristic: there is no *resupination* (the ovary is not twisted 180° round its axis), so the lip is pointing upwards between the paired sepals which together form a typical, helmet-like structure, see Fig. 4d. The flowering stalk arises from the top of the mature pseudobulb and is usually covered by leaves at least in its lower part. The leaves are generally thin and often deciduous. Most Polystachyas are easily cultivated but of limited interest because of the small-sized and dull-coloured flowers in many species.

Polystachya laxiflora Lindl. [Fig. 61]

A sturdy epiphyte with long, cylindrical pseudobulbs, 8-10 cm long, 0.4-0.9 cm wide, and rather large, yellow flowers.

Leaves: Oblanceolate-elliptic-acute, 80-250 mm long, 20-55 mm wide, apex curved slightly backward.

Dorsal sepal: Triangular-acuminate, 7-10 mm long, 5 mm wide, pale yellow, hairy.

Lateral sepals: Triangular-acuminate, 12 mm long, 7 mm wide, pale yellow, hairy.

Petals: Spatulate-obtuse, 6 mm long, 3 mm wide, pale yellow, glabrous.

Lip: Trilobed, 10 mm long. Middle lobe triangular-acute, yellow with purple markings, pubescent, and with a callus 2 mm long, 1 mm wide. Side lobes rounded, pale yellow with purple streaks.

Distribution: S.

Polystachya saccata (Finet) Rolfe. [Fig. 62]

A small epiphyte, 8-10 cm high, leafless at time of flowering. Pseudobulbs broadly ovate to globular, 0.5 cm wide. Scape with dense white hairs. Inflorescences many-flowered.

Bracts: Triangular-acuminate, 2 mm long.

Dorsal sepal: Oblong-apiculate, 2.5 mm long, 1.2 mm wide, pale lilac.

Lateral sepals: Broadly triangular-apiculate, often much recurved, 3 mm long, pale lilac.

Petals: Broadly spatulate, 2.5 mm long, 1 mm wide, very pale lilac.

Lip: Oblanceolate-obtuse, 4 mm long, with sac at apex; very sharply reflexed below the middle. Sac with large yellow hair cushion. Apex of lip yellow, base pale yellow to pale lilac.

Column: Pale green, operculum pale lilac.

Distribution: N.

Polystachya coriscensis Rchb.f. [Fig. 63 & 64]

A small, short-stemmed epiphyte, 5-12 cm high, with leaves tufted together at base of stem. The leaf stalks are covered with large reticulate sheaths. The plant has numerous roots, some of which are bright purple.

Leaves: Elliptic, with slightly unequally and obtusely bilobed apex, blade 45-55 mm long, 20-22 mm wide. Stalk 15 mm long.

Bracts: Acuminate, clasping, 6 mm long, 2.5 mm wide, green.

Dorsal sepal: Ovate-acuminate, 3.0-4.5 mm long, 2 mm wide, green with purple margins and two diffuse, parallel purple lines in centre.

Lateral sepals: Triangular-acuminate, 5 mm long, 4 mm wide, green, with slight purple markings on margins.

Fig. 65. *Polystachya talbotii.* In montane rain forest, altitude 1 600 m, Obudu, Cross River St. October. 5x.

Fig. 66. *Polystachya adansoniae.* Uhoko Forest, Imo St. June. 1.8x.

Fig. 67. *Polystachya odorata.* 3x.

Petals: Linear-acute, slightly widened at apex, 3 mm long, 5 mm wide, pale green with purple margins in lower half.
Lip: Trilobed, 4 mm long, 2 mm wide. Base of lip rectangular, pubescent, white. Middle lobe ovate-acute, much reflexed, 1.5 mm long, pale yellowish green. Side lobes rounded, incurved, purple.
Distribution: S.

Polystachya talbotii Rolfe. [Fig. 65]
A small, but very attractive epiphyte with ovoid pseudobulbs and rather large, pink flowers. It is a montane species and, like many other high-altitude epiphytes, not easy to cultivate. The inflorescence is branched and many-flowered. The slender flowering stalk has tightly clasping, dark brown sheaths. This species is considered to be identical with *Polystachya alpina* Lindl.
Leaves: Linear-lanceolate, 50-100 mm long, 6-8 mm wide, unequally and obtusely bilobed at apex.
Dorsal sepal: Lanceolate-acute, 8 mm long, 4 mm wide, very pale pink.
Lateral sepals: Ovate-acute or apiculate, close together at base, 10 mm long, 4 mm wide, very pale pink.
Petals: Oblanceolate-acute, 5 mm long, 1.5 mm wide, very pale pink.
Lip: Ovate-apiculate, apex recurved, cream-coloured or very pale pink, with large, fleshy, four-lobed callus, wider than blade of lip. Lobes erect, fleshy, cream-coloured with pink edges.
Column: Dark purple (including operculum).
Distribution: S.

Polystachya adansoniae Rchb.f. [Fig. 66]
An epiphyte with narrowly conical pseudobulbs, which are 1-6 cm high and 0.5-0.9 cm wide. The plant is often found on Baobab trees. Its white or yellow flowers are faintly scented. The inflorescence is dense, erect, unbranched and many-flowered, 5-14 cm long, and has a bristly appearance owing to the long drawn-out points of the sepals.
Leaves: Linear-lanceolate, thin, apex very indistinctly and obtusely bilobed, blade often folded lengthwise, 40-250 mm long, 4-20 mm wide.
Bracts: Subulate with narrow base, 6-14 mm long, 1 mm wide.
Dorsal sepal: Triangular-acuminate, 3-4 mm long, 1.5 mm wide, greenish white to yellow.
Lateral sepals: Triangular-acuminate, 5 mm long, 3 mm wide, greenish white to yellow.
Petals: Linear-acute, 2.5 mm long, 0.3 mm wide, greenish white to yellow.
Lip: Trilobed, 3 mm long, 2 mm wide, greenish white to yellow. Middle lobe triangular-acuminate, much reflexed. Side lobes rounded, erect.
Column: Dark purple (including operculum).
Distribution: N + S.

Polystachya odorata Lindl. [Fig. 67 & 68]
A variable species, with a wide distribution in Nigeria. The inflorescence is much branched, and many-flowered. Rhachis, bracts, ovary and sepals are covered with sparse fine hairs. In spite of the specific name, the author has never found a scented specimen of this orchid.

Fig. 68. *Polystachya odorata.* Owerri, Imo St. April-June, September. 0.4x.

Fig. 69. *Polystachya tessellata.* Uhoko Forest, Imo St. June-July. 1.8x.

Fig. 70. *Polystachya modesta.* Savanna, Kufana, Kaduna St. September. 0.6x.

Fig. 71. *Polystachya modesta.* 4.5x.

Leaves: Lanceolate-oblong-acute, 150-300 mm long, 25-35 mm wide; lower leaves ovate, smaller.
Bracts: Caudate, reflexed, 3-6 mm long, 2-2.5 mm wide.
Dorsal sepal: Ovate-acute, 4-6 mm long, 4 mm wide.
Lateral sepals: Triangular-acuminate, 9 mm long, 5 mm wide. All sepals pale green with purple markings on base.
Petals: Linear-acute, slightly spatulate, 5-6 mm long, 1 mm wide, very pale green, base purple.
Lip: Trilobed, 6-8 mm long. Lower part of lip white, with purple base and distinct keel, widening to yellow callus, 8 mm wide. Middle lobe orbicular-acute, seemingly emarginate because of much reflexed apex, 4 mm long and wide, pale greenish yellow, with irregular margins. Side lobes triangular, erect, white.
Distribution: N + S.

Polystachya tessellata Lindl. [Fig. 69]

A common and variable species, up to 50 cm high, but usually smaller. The pseudobulbs are closely placed, conical, 0.8 cm high, 1 cm wide. Inflorescence 25-35 cm high, 8-14-branched, nearly secund, distance between branches 2-2.5 cm. Rhachis with sparse fine hairs and thin, dry sheaths. Faintly scented.
Leaves: Oblanceolate (or oblong, elliptical), unequally and acutely bilobed at apex, 60-300 mm long, 15-60 mm wide, dark green with purple stains, especially on lower side.
Dorsal sepal: Triangular-apiculate, 2.5-4.5 mm long, 2 mm wide, pale green, apex purple.
Lateral sepals: Triangular-apiculate, 5 mm long, 4 mm wide, pale green with purple apex. Sometimes with purple 'chessboard' pattern = tessellated.
Petals: Linear, with widened, acute apex, pale green.
Lip: Trilobed, 4-5 mm long. Middle lobe broadly triangular, irregularly dentate and indistinctly retuse at apex, 3 mm wide, with brick-coloured pubescent callus, forming part of pubescent keel which extends towards base of lip. Middle lobe white to pale violet, with wavy margins. Side lobes entire, acute, slightly incurved.
Distribution: S (+ N).

Polystachya modesta Rchb.f. [Fig. 70 & 71]

Epiphyte with small, conical pseudobulbs. Flowering stalk 10-25 cm long, 3-6-branched, lower part with dry, apiculate sheaths. Branches 1.5-2.0 cm long, not much spreading. Inflorescences many-flowered, with branches pointing in all directions. The flowers are fleshy and greenish white, turning yellow. Faintly fragrant.
Leaves: Linear-oblong (or narrowly oblanceolate), 60-110 mm long, 6-22 mm wide, very slightly and acutely bilobed at apex, thin.
Bracts: Ovate-acute, somewhat thickened, with sparse fine hairs in lower part of inflorescence, 3 mm long, 1.5-2.0 mm wide, yellowish green.
Dorsal sepal: Ovate-obtuse, slightly incurved, 2.5-3.5 mm long, 2 mm wide.
Lateral sepals: Triangular-apiculate, 5 mm long, 3 mm wide.
Petals: Linear-obtuse, slightly incurved, 2 mm long, 0.6 mm wide.

Fig. 72. *Polystachya golungensis.* In gallery forest, Zutung River, Kaduna St. January. 0.7x.

Fig. 73. *Polystachya golungensis.* 9x.

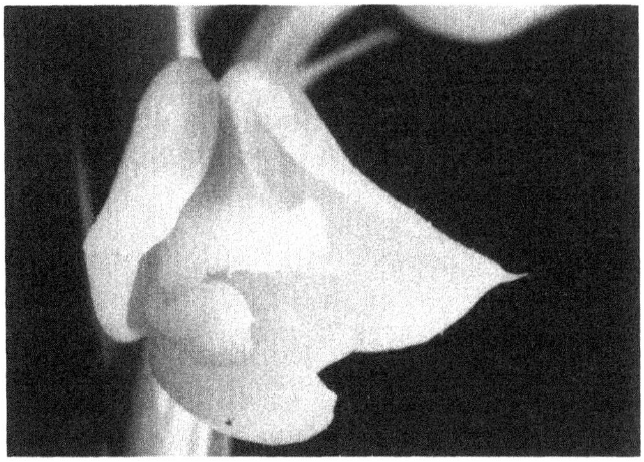

Fig. 75. *Polystachya subulata.* 7.5x.

Fig. 74. *Polystachya subulata.* Savanna, Anara F.R., Kaduna St. May. 0.6x.

Lip: Trilobed, 3-5 mm long, 2 mm wide. Blade bullate, base sparsely, middle part densely pubescent, concave. Middle lobe triangular, recurved, with crenate margin, no callus. Side lobes small, erect, rounded triangular.
Distribution: N + S.

Polystachya golungensis Rchb.f. [Plate 2.1; Fig. 72 & 73]
Epiphyte with very closely placed, elongate-conical, somewhat laterally flattened pseudobulbs, 1-7 cm high, 0.8-1.5 cm wide, often with slight bend.
Leaves: 2-3, linear-oblong, obtusely and unequally bilobed at apex, very thick, broadly V-shaped in section, 35-200 mm long, 11-25 mm wide, yellowish green, edges sometimes tinged purple.
Flowering stalk: 100-300 mm high, branched, covered with sheaths. In small individuals, side branches are often represented by arrested buds in axils of stem bracts. Side branches up to seven, more or less erect and secund, 15-40 mm long.
Rhachis: Widening towards apex of inflorescence, densely many-flowered and with sparse fine hairs.
Bracts: Small, triangular-acuminate, hairless, 1.5-2.0 mm long, 1 mm wide.
Flowers: Yellowish green, fleshy, faintly and unpleasantly scented.
Dorsal sepal: Ovate-apiculate (or acute), 2 mm long, 1-1.5 mm wide.
Lateral sepals: Obliquely triangular-apiculate (or acute), 2.5-3.5 mm long, 2.5 mm wide.
Petals: Linear-subacute, very slightly widened near apex, 1.5-2.0 mm long, 0.3-0.5 mm wide.
Lip: Very distinctly trilobed, 3-5 mm long. Middle lobe suborbicular-emarginate, 2 mm Ø, with wavy margin, apex pointing upwards. Side lobes obliquely obtuse, erect, 1.5 mm long. Basal half of lip with white, densely pubescent cushion in centre.
Seed capsules: 4-5 mm long.
Distribution: N + S.

Polystachya subulata Finet. [Fig. 74 & 75]
Small epiphyte with conical, often purple-coloured pseudobulbs, up to 0.7 cm wide. The inflorescence is branched, the rhachis has sparse fine hairs and white, pink or purple flowers.
Leaves: Narrowly lanceolate-acute, slightly bilobed at apex, 50-200 mm long, 8-30 mm wide, green, tinged purple in purple-flowered individuals.
Bracts: Acuminate, with a wide base, 5 mm long.
Sepals: Triangular-acuminate, 4-5.5 mm long, 2.5 mm wide (dorsal), 7 mm long, 4 mm wide (laterals).
Petals: Spatulate-apiculate, 4 mm long, 1.2 mm wide, white or cream-coloured.
Lip: Very indistinctly trilobed, cordate-acuminate with short hairs; 5 mm long, 4 mm wide, white or very pale green. Pale mauve or purple callus in lower half.
Distribution: N.

Polystachya affinis Lindl. [Fig. 76 & 77]
An attractive, rather large-flowered epiphyte with hanging habit. Pseudo-

Fig. 76. *Polystachya affinis.* In rain forest, Ubong, Cross R. St. January.

Fig. 77. *Polystachya affinis.* 4.5x.

Fig. 78. *Ancistrochilus rothschildianus.* Forest, Benin, Bendel St. November. 0.9x.

Fig. 78a. *Ancistrochilus rothschildianus.* 1.8x.

PLATE 3

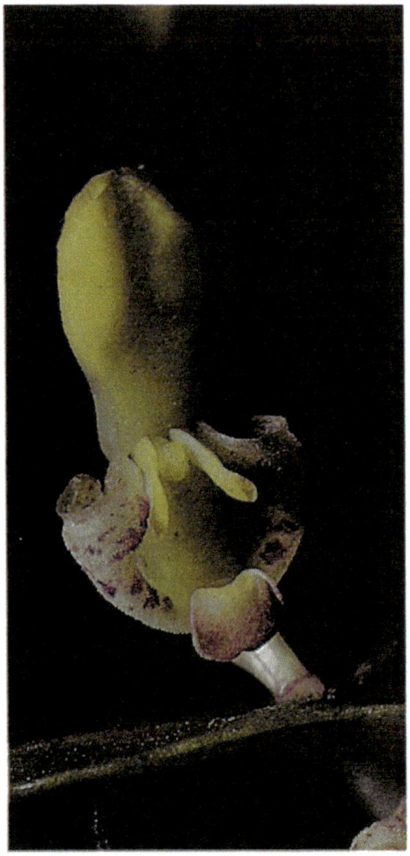

bulbs very characteristic: orbicular or ovate, much flattened, pressed close to the bark of the host, diameter 2.5-5 cm, three-leaved. Inflorescence 12-21 cm long, with a few branches, 4-5 cm long. Peduncle covered with dry sheaths. Whole inflorescence covered with fine hairs, 10-30-flowered, not scented.

Leaves: Oblanceolate-acuminate. blade 80-160 mm long, 30-45 mm wide, petiole 30 mm.

Bracts: Large, ovate-lanceolate-acuminate, 10-12 mm long, 5-6 mm wide.

Dorsal sepal: Obtuse, apex recurved, 7 mm long, 4 mm wide, pale yellow.

Lateral sepals: Acute or obtuse, apex recurved, 8 mm long, 7 mm wide, pale yellow with purple veins.

Petals: Oblanceolate, converging, 6-7 mm long, 2 mm wide, pale yellow.

Lip: Indistinctly trilobed, 7-8 mm long, 5-6 mm wide, with long keel in lower part. Sides of keel pubescent. Middle lobe obtuse, reflexed, side lobes erect. Whole lip pale yellow with purple veins.

Distribution: S.

GENUS ANCISTROCHILUS

This genus of handsome but rather uncommon rain forest epiphytes has two members in Nigeria. The flowers are quite large, the diameter being often more than 5 cm. They are attractively coloured in shades of mauve and purple. The middle lobe of the lip is long, sharply pointed and has a deeper colour than the rest of the flower.

Ancistrochilus rothschildianus O'Brien [Cover; Fig. 78]

A medium-sized epiphyte with large (5-6 cm wide), rose to mauve-coloured flowers. The pseudobulbs are characteristically pear- or onion-shaped, and up to 5 cm wide. The two thin, plicate leaves arise from the top of the pseudobulb, the pubescent flowering stalk from its base.

Leaves: Elliptical-lanceolate, 100-200 mm long, 50 mm wide.

Dorsal sepal: Oblanceolate-obtuse, 20-30 mm long, 10 mm wide, pink.

Lateral sepals: Oblanceolate-obtuse, 20-30 mm long, 9 mm wide, pink.

Petals: Oblanceolate-obtuse, 20-30 mm long, 6 mm wide, pink.

Lip: Trilobed, 15 mm long. Middle lobe narrowly triangular, recurved, 10 mm long, 4 mm wide, purple. Side lobes rounded, erect, greenish or brownish.

Column: Long, slender, slightly incurved.

Distribution: S.

Plate 3

Above left: ***Bulbophyllum schinzianum*** Kränzl. 5.1x.

Above centre: ***Bulbophyllum falcatum*** (Lindl.) Reichb.f. 9.6x.

Above right: ***Bulbophyllum calyptratum*** Kränzl. 8x.

Below left: ***Bulbophyllum maximum*** (Lindl.) Reichb.f. 7x.

Below right: ***Bulbophyllum distans*** Lindl. 4.5x.

Fig. 80. *Bulbophyllum flavidum.* 5x.

Fig. 79. *Bulbophyllum schinzianum.* 3.5x.

Fig. 82. *Bulbophyllum distans.* On *Chlorophora* sp., Ukogugu Forest, Imo State. April. 3x.

Fig. 81. *Bulbophyllum flavidum.* In forest, Benin, Bendel St. July. 1x.

GENUS BULBOPHYLLUM

This is a very large and important genus of epiphytic orchids with a pan-tropical distribution – there are more than 1 000 species. The plants have well-defined 1- or 2-leaved (very rarely 3-leaved) pseudobulbs. The leaves are generally thick and persistent and arise from the top of the pseudobulb, whereas the flowering stalk is placed at its base.

The flowers are generally quite small, but well worth studying in detail, as they are very unusual in appearance, especially in the purely African section of the genus called 'Megaclinium', see Fig. 4g. In this section, the flowers alternate on both sides of a thickened and/or flattened, often leaf-like rhachis.

Typically, the lip is very loosely attached with a narrow 'claw' to the base of the column, and the smallest breath of air will make it move: perhaps a way of attracting insects. The whole flower is often quite insect-like, with petals resembling antennae and a lip that is patterned or furry in the same way some insects are.

Bulbophyllum flavidum Lindl. [Fig. 80 & 81]

A medium-sized epiphyte. The flowering stem is slender and rounded, not thickened, and 7-18 cm high, with conical, 1-leaved pseudobulbs, 1-3 cm high, 0.7-1.0 cm wide.

Leaves: Narrowly oblong-acute, 70-140 mm long, 7-30 mm wide.

Bracts: Lanceolate-acute, 3 mm long, 1 mm wide.

Sepals: Lanceolate-acute, 7 mm long, 2 mm wide, yellowish green with purple base.

Petals: Obovate-obtuse, 2 mm long, 1 mm wide, white.

Lip: Tongue-like, apex recurved, 1.5 mm long, 0.7 mm wide, yellow.

Distribution: S.

Bulbophyllum distans Lindl. [Plate 3.5; Fig. 82 & 83]

A large species with 1-leaved pseudobulbs closely placed on raceme: spacing 2 cm, or less. Pseudobulbs lenticular with one, rather indistinct ridge on one of the two sides; 2.3-3.5 cm high, 2-3 cm wide, yellowish green. Peduncle 35 cm long, 0.2 cm wide, inflorescence 25-30-flowered, rhachis not thickened.

Leaves: Narrowly oblong-subacute, or very slightly bilobed at recurved apex, leathery, blade 80-130 mm long, 20 mm wide, petiole 10 mm long.

Bracts: Lanceolate-acute, scarious, 8-10 mm long, 1.2-2.5 mm wide.

Sepals: Lanceolate-acute, slightly recurved, 9-11 mm long, 2 mm wide, yellow-green.

Petals: Subulate, 3-4 mm long, 0.5 mm wide, brown.

Lip: Linear-acute, 7-9 mm long, 2 mm wide at base, with reddish brown, 4 mm long hairs.

Column: With subulate stelidia.

Distribution: S.

Bulbophyllum schinzianum Kraenzl. [Plate 3.1; Fig. 79 & 84]

A sturdy epiphyte with closely placed, 1-leaved, prismatic-ovoid, 4-(3-)-ridged pseudobulbs, 2-4 cm high, 1.5-3.5 cm wide. The flowering stalk is

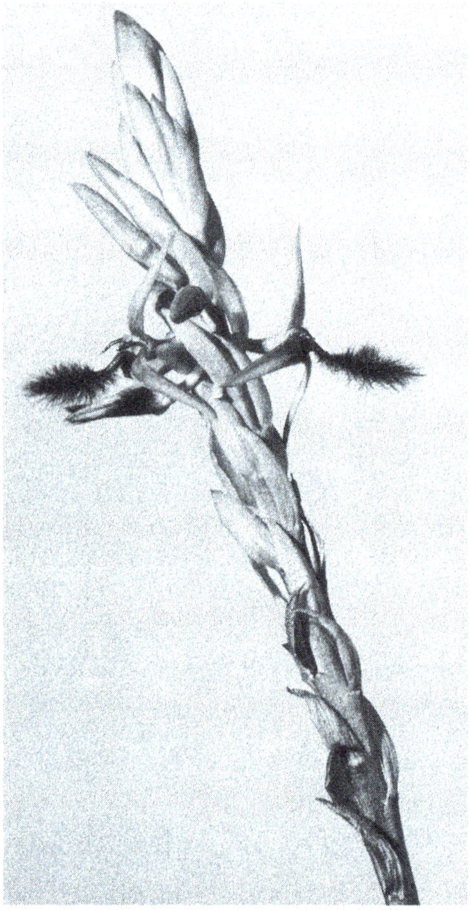

Fig. 83. *Bulbophyllum distans.* 1.5x.

Fig. 86. *Bulbophyllum oreonastes.* 1.4x.

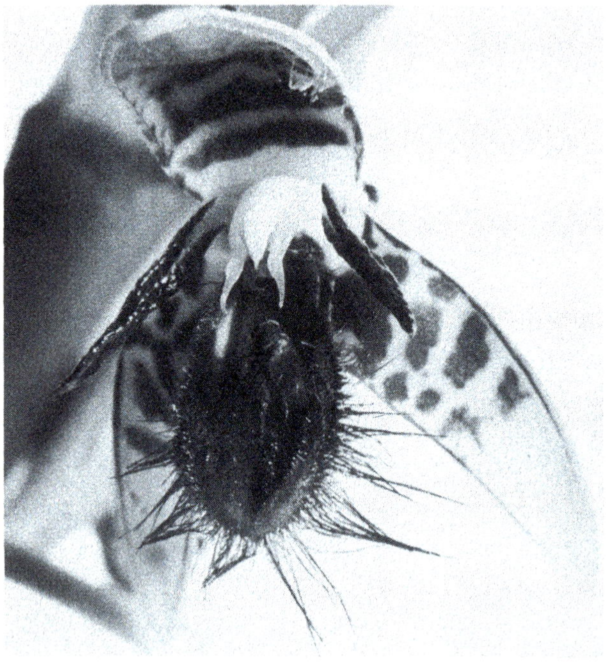

Fig. 84. *Bulbophyllum schinzianum.* On *Chlorophora excelsa,* Ukogugu Forest, Imo St. March. 7.5x.

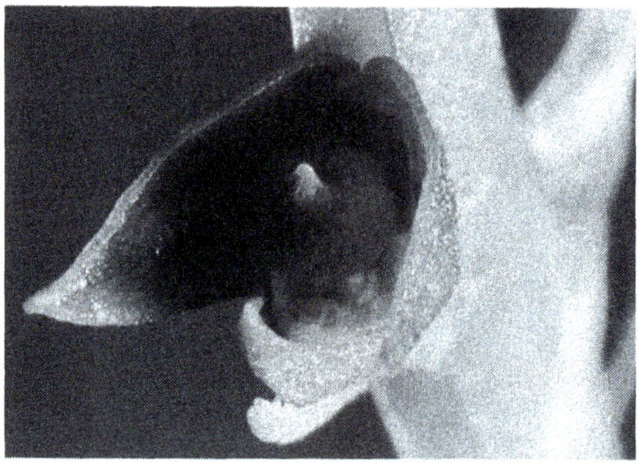

up to 55 cm long, 0.2 cm wide; the rhachis is 15 cm long and not flattened. The outer side of the sepals is provided with nectaria which are much frequented by ants. The flowers are sweetly scented in day-time.

Leaves: Oblong-obtuse, thick and leathery, very slightly bilobed at apex, 100-200 mm long, 25-50 mm wide, petiole about 10 mm long.

Bracts: Elliptic-acute, thin and dry, 20 mm long, 8 mm wide.

Dorsal sepal: Broadly lanceolate-acute, 12 mm long, 5 mm wide, pale yellow with four purple streaks across the lower half; upper half with faint purple spots.

Lateral sepals: Lanceolate-acute, slightly falcate, 9 mm long, 3 mm wide, pale greenish yellow, with purple streaks or flecks in basal half.

Petals: With lanceolate-acute blade, 5.5 mm long, 1 mm wide, dark brown, and linear stalk, 3 mm long, 0.5 mm wide, pale green.

Lip: Slightly thickened, with narrow (0.1 mm wide) linear claw and elliptical blade. Length of lip 8 mm, width of blade 3.5 mm. Hypochilium pectinate, with two white calli. Mesochilium olive-green with two spreading keels that disappear towards centre of blade, edges thickened, greyish brown, puberulent. Epichilium black with single keel in middle, apex light brown, margin pubescent, underside of margin bearded, hairs deflexed, up to 4 mm long. Lower side of lip dark brown, scabrous.

Column: With two acuminate, hyaline stelidia, 2.5 mm long, 0.5 mm wide. Base of column pale green, with rectangular purple field.

Distribution: N + S.

Bulbophyllum oreonastes Rchb.f. [Fig. 85 & 86]

A small epiphyte with parallelepipedic-conical pseudobulbs, 1.8-3 cm high, 0.7-1 cm wide, 2-leaved with four angles. Rhizome 0.2 cm thick, pseudobulbs spaced 2.5-7 cm. Scape 5-10 cm long, peduncle 0.1 cm wide. Rhachis somewhat flattened, canaliculate, 5-6.5 cm long, 0.25-0.3 cm wide, 0.1-0.15 cm thick, 9-13-flowered.

Leaves: Oblong-elliptical-obtuse, 30-70 mm long, 8-14 mm wide, slightly and unequally bilobed at apex.

Bracts: Acuminate, 4 mm long, 2 mm wide, pale yellow.

Dorsal sepal: Lanceolate-acute, orange-brown (or greenish yellow with purple base), bent down over column and petals, 4 mm long, 1.5-2 mm wide.

Lateral sepals: Obliquely triangular-apiculate, 4 mm long, 2 mm wide, orange-brown (or greenish yellow with purple base).

Petals: Linear-acute, slightly reflexed and swollen at apex, 2-5 mm long, 0.3-1.0 mm wide, mauve or orange-brown, with cream or yellow apex.

Lip: Tongue-like, much swollen, fleshy, very loosely attached to column, 2.5-4.0 mm long, 1.5-2.5 mm wide, 1 mm thick, purplish red or brownish purple.

Rostellum: Winged, pale purple.

Distribution: N + S.

Fig. 85 (centre right). *Bulbophyllum oreonastes.* Ukogugu Forest, Imo St. Januari. 13x.

Fig. 87 (below right). *Bulbophyllum zenkeranum.* Ukogugu Forest, Imo St. April. 1x.

Bulbophyllum zenkeranum Kraenzl. [Fig. 87]

Very small epiphyte with narrowly ovoid, 2-leaved pseudobulbs, 0.7-1.2 cm high, 0.3-0.7 cm wide, spaced 1.2-1.5 cm along a thin rhizome, 0.1 cm wide. Pseudobulbs in old, not flowering individuals with three marked ridges. Scape 2.5-3.0 cm high, lower part and pseudobulb covered

Fig. 88. *Bulbophyllum lupulinum.* Gallery forest, Zutung, Kaduna St. February. 0.7x.

Fig. 89. *Bulbophyllum lupulinum.* 3.6x.

by sheath. Rhachis somewhat thickened and flattened, canaliculate, with 3-5 flowers set in the grooves.
Leaves: Elliptical, very slightly unequally bilobed at apex, 12-20 mm long, 4-6 mm wide.
Bracts: Lanceolate-acuminate, 4-5 mm long.
Sepals: Lanceolate-acuminate, 5 mm long, 2 mm wide, yellow, with brown longitudinal streaks.
Petals: Linear-obtuse, 1.5 mm long, 0.3 mm wide, white.
Lip: Lanceolate-obtuse, 1.5 mm long, 0.3 mm wide, white.
Distribution: S (+ N).

Bulbophyllum lupulinum Lindl. [Fig. 88 & 89]
Sturdy epiphyte with 2-leaved, elliptical, sharply 4-winged pseudobulbs, 5-6 cm high.
Leaves: Oblong-acute, or indistinctly and obtusely bilobed at apex, 90-120 mm long, 20-30 mm wide, thick, slightly arching.
Peduncle: 4 mm Ø, 200 mm long, with large triangular sheath at base; the tightly clasping sheaths on the stalk have minute brown spots.
Rhachis: Pale (bluish-)green, with small brown spots (scurfy), intermittently canaliculate with flowers in the grooves. Pedicels very short.
Bracts: Very large, almost covering flowers, triangular-acute, 15 mm long, 8 mm wide, scarious, light brown with darker brown spots.
Dorsal sepal: Lanceolate-apiculate, 5 mm long, 2 mm wide, slightly incurved, dorsal side pale yellow with purple spots, ventral side pale yellow.
Lateral sepals: Obliquely lanceolate-apiculate, 4 mm long, 2 mm wide, pale yellow with purple spots, very slightly recurved.
Petals: Linear-acute, 3 mm long, 0.3 mm wide, placed close to column, slightly arching, pale yellow with very small purple spots.
Lip: Tongue-like, motile, trilobed, 2 mm long, 1 mm wide, pale yellow with purple pattern. Side lobes elongate, slightly rounded, erect. Middle lobe shorter, rounded, recurved. Claw narrow, very short.
Column: Pale purple with purple markings. Foot of column with very dark purple pattern. Operculum pale yellow.
Distribution: N + S.

Bulbophyllum congolanum Schltr. [Plate 4.2; Fig. 90 & 91]
Large epiphyte (scape up to 65 cm long) with conical, 2-(3-)leaved pseudobulbs with (3)-4 angles, 2.5-6 cm high, 1.5-2.5 cm wide. Rhachis cream-coloured (or green) with numerous purple spots, or entirely purple (purplish brown), thickened and flattened, pointed at apex, 12-22 cm long, 0.9-1.0 cm wide, 0.2-0.4 cm thick.
Leaves: Lanceolate-acute, thick and leathery 60-190 mm long, 15-30 mm wide.
Dorsal sepal: Narrowly lanceolate-acute, 6-8 mm long, outer side purplish brown, inner side cream-coloured with purple streaks.
Lateral sepals: Falcate, acuminate, cream-coloured with irregular purple markings.
Petals: Narrowly linear, 3-5 mm long, purplish brown.
Lip: Ovate-obtuse, much deflexed, cream-coloured with purple markings.

Fig. 90. *Bulbophyllum congolanum.* On *Pentaclethra* sp., Oguta, Imo St. October. 8x.

Fig. 91. *Bulbophyllum congolanum.* 2.5x.

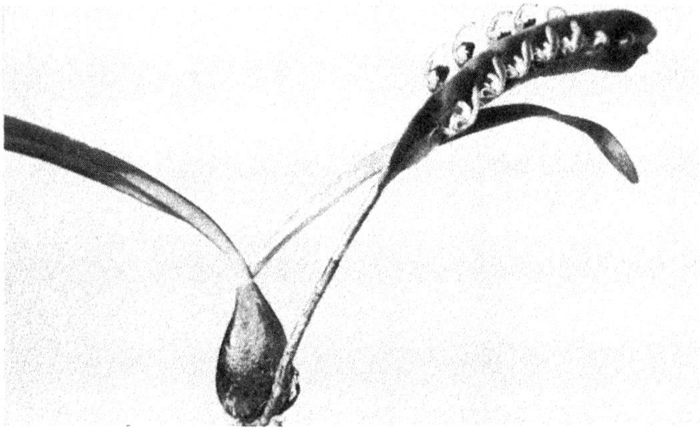

Fig. 92. *Bulbophyllum falcatum.* Forest, Ohusu, Bendel St. October. 0.9x.

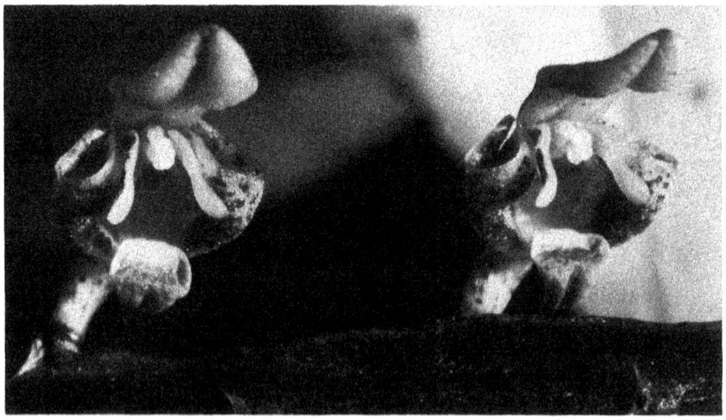

Fig. 93. *Bulbophyllum falcatum.* 7.5x.

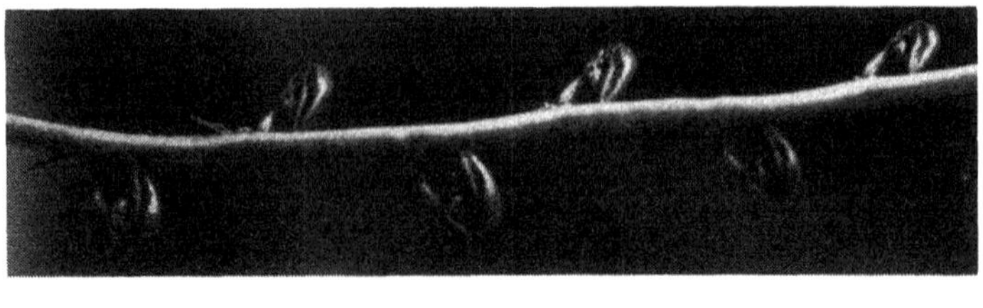

Fig. 94. *Bulbophyllum calyptratum.* 1.5x.

Column: Purple with cream-coloured rectangular field below stigma. Stelidia rounded triangular, cream-coloured. Operculum yellow.
Distribution: N + S.

Bulbophyllum falcatum (Lindl.) Rchb.f. [Plate 3.2; Fig. 92 & 93]
Pseudobulbs ovoid, 3-4-(5-)angled, 1.5-5 cm high, 1.0-1.8 cm wide. A typical Megaclinium-type, with thin, flattened rhachis up to 1.5 cm wide, green with purple spots.
Leaves: Oblanceolate, unequally and obtusely bilobed at apex, 40-150 mm long, 6-20 mm wide.
Bracts: Lanceolate-acute, 2-2.5 mm long.
Dorsal sepal: Spatulate with thickened margins, 5-7 mm long, 2 mm wide, outer side purple, inner side white, margins yellow.
Lateral sepals: Lanceolate-acute, much reflexed, 3 mm long, 2.5 mm wide, white with purple markings.
Petals: Narrow, club-like, 2 mm long, 0.2 mm wide, white with yellow tips.
Lip: Small, tongue-like, much recurved, purple.
Distribution: S.

Bulbophyllum calyptratum Kraenzl. [Plate 3.3; Fig. 94 & 95]
Pseudobulbs conical, 2-leaved, 3-angled, 5-7 cm high, 1.7-2.3 cm wide. Scape 35-50 cm long, peduncle 0.2 cm wide. Rhachis flattened, 20-30 cm long, 0.3-0.4 cm wide. Fully opened flowers reflexed, making inflorescence nearly secund.
Leaves: Linear, slightly bilobed at apex, 150-190 mm long, 9-12 mm wide.
Bracts: Triangular, reflexed, 5-6 mm long, 4 mm wide.
Dorsal sepal: Broadly spatulate, 6-9 mm long, 3 mm wide, upper side dark purple, lower side pale green with purple spots.
Lateral sepals: Acuminate, much reflexed, 6 mm long, pale green with dark purple spots.
Petals: Oblong-lanceolate-acute, 1.5-2.0 mm long, 0.3-0.5 mm wide, slightly thickened at apex, pale greenish yellow.
Lip: Narrowly tongue-like, much recurved, 2 mm long, 0.6 mm wide, purple.
Distribution: S.

Bulbophyllum oxypterum (Lindl.) Rchb.f. [Fig. 96]
Rhizome light brown, scaly, 0.35 cm Ø, distance between pseudobulbs 5 cm. Pseudobulbs elongate-conical, slightly recurved and contorted, 2-leaved, 3-4-angled with additional sharp ridges, 3-5.5 cm high, 1 cm wide. Lower part of pseudobulb with tightly clasping sheaths. Peduncle with tightly clasping sheaths, 12-19 cm long, 0.2-0.3 cm wide, lower part yellowish green, gradually changing to dark purple brown (spotted). Rhachis making blunt angle with peduncle, or peduncle recurved. Rhachis flattened, 70-90-flowered, linear-acuminate, 17-25 cm long, 0.8-1.0 cm wide, 0.30-0.35 cm thick, lower part cream-coloured with minute dark brown spots, upper part purplish dark brown. Flowers opposite, only two open at any time.
Leaves: Oblong-lanceolate, obtuse or very slightly bilobed at recurved

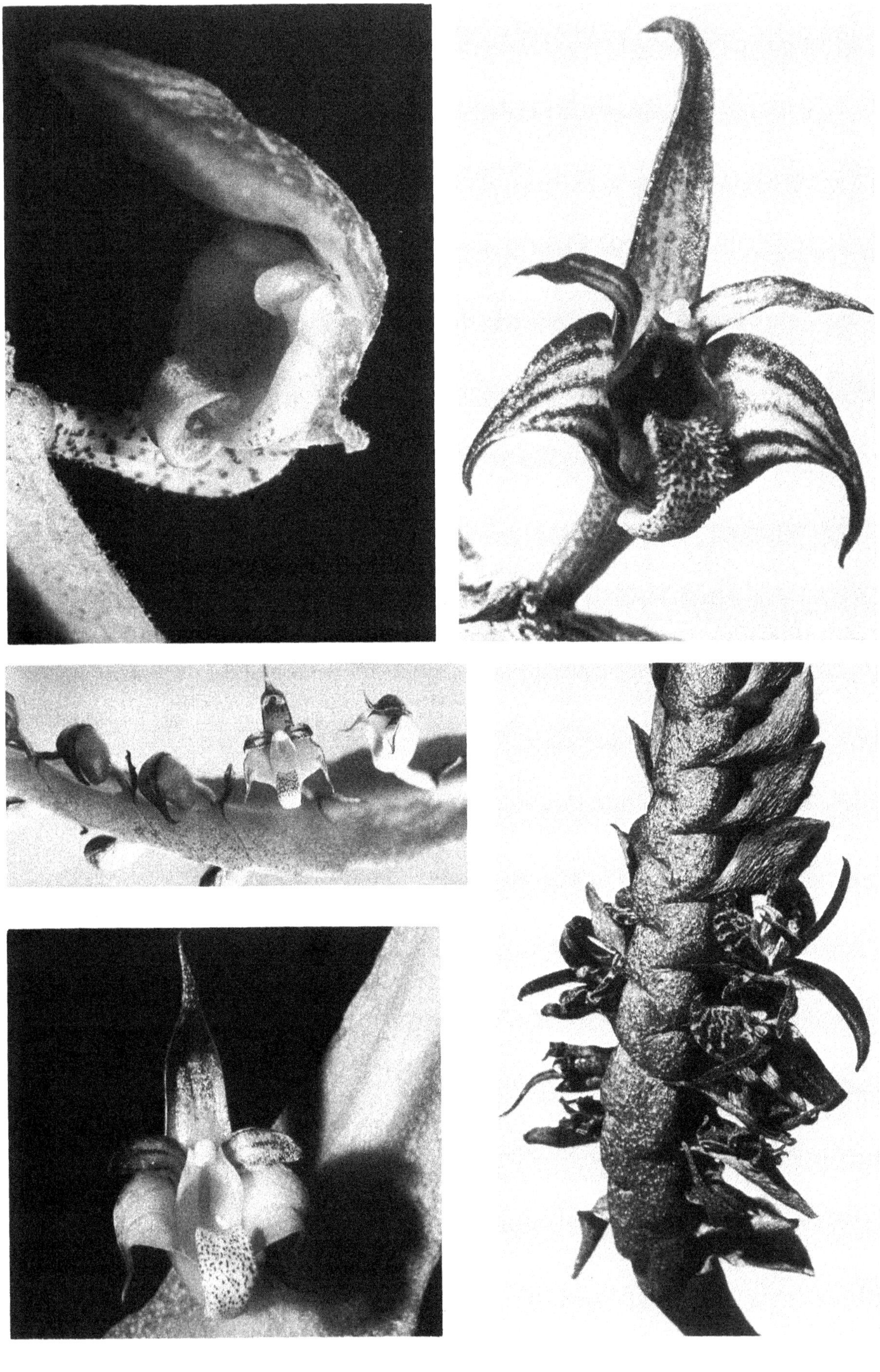

apex, 75-150 mm long, 20-40 mm wide, dark green, fleshy. Middle vein seen as distinct keel on back of leaf.
Bracts: Triangular, fully withered when flower opens, 3 mm long, 3 mm wide, at base.
Dorsal sepal: Lanceolate-acuminate, 7 mm long, 1 mm wide, purple or cream-coloured with purple spots; apex dark purple, recurved.
Lateral sepals: Triangular or nearly orbicular, acuminate, apex recurved, 5 mm long, 3 mm wide, base light purple, middle part and apex cream-coloured with three purple stripes, edges purple.
Lip: Linear-obtuse, slightly swollen, recurved, 2.5 mm long, 0.8 mm wide. Margins of hypochilium pectinate. Upper side of lip cream-coloured with purple spots, mesochilium and epichilium of lower side light yellow.
Column: With purple, triangular wings.
Distribution: N + S.

Bulbophyllum maximum (Lindl.) Rchb.f.* [Plate 3.4; Fig. 97 & 98]
Pseudobulbs ellipsoid, 1-2-leaved, (3-)4-ridged, 3.5-5 cm high, 1.5-2 cm wide, spaced 4-5 cm on raceme 0.3-0.4 cm wide. Scape 25-35 mm high, with brown sheathing bracts on peduncle. Rhachis flattened and pointed, 10-15 cm long, 1.2-1.5 cm wide, yellowish green with brown spots on edges. Flowers unpleasantly scented (like *Castanea sativa*).
Leaves: Oblong-obtuse, thick, very slightly bilobed at apex, 80-100 mm long, 22-30 mm wide.
Dorsal sepal: Acuminate, 12 mm long, 3 mm wide, apex dark brown, base pale yellow.
Lateral sepals: Acuminate, much reflexed, pale yellow, apex dark brown.
Petals: Oblong-acuminate, slightly recurved, 4 mm long, pale yellow with dark brown line in centre, margins dark brown.
Lip: Tongue-like, swollen, much recurved, 7 mm long, pectinate in basal half; pale yellow with small purple dots.
Column: Pale yellow, operculum yellow.
Distribution: S.

Bulbophyllum kamerunense Schltr. [Fig. 99]
Pseudobulbs ovoid, 3-(4-)ridged, 3.5-5 cm high, 1.5-3.5 cm wide, 2-(3-)-leaved. Rhizome 0.6-0.7 cm wide. Scape 30-50 cm long, peduncle 0.35-0.40 cm wide, with 13-15 sheathing bracts. Rhachis 10-18 cm long, 0.7-1 cm wide, 0.5 cm thick, 30-40-flowered, only 3 or 4 flowers open at any time. Sweetly scented in day-time.
Leaves: Oblong-lanceolate, 80-170 mm long, 15-35 mm wide, thick and leathery.
Bracts: Triangular-obtuse, 12 mm long, 7 mm wide, brown. Young bracts with nectaria.
Dorsal sepal: Lanceolate-acute, recurved, with dense and fine hairs, 5-12 mm long, 2.5 mm wide, cream-coloured with dark purple, irregular patches, or deep purple with white base.

* Not identical with *B.maximum* (Ridl.) Ridl. 1907 (= *Cirrhopetalum maximum* Ridl. 1900), a Malaysian species.

Fig. 95 (above left). *Bulbophyllum calyptratum.* Forest, Okigwi, Imo St. November. 11x.

Fig. 96 (above right). *Bulbophyllum oxypterum.* Ukogugu Forest, Imo St. May. 13x.

Fig. 97 (centre left). *Bulbophyllum maximum.* Ukogugu Forest, Imo St. March. 1.5x.

Fig. 98 (below left). *Bulbophyllum maximum.* 4.8x.

Fig. 99 (below right). *Bulbophyllum kamerunense.* On *Pentaclethra* sp., Oguta, Imo St. February. 3.5x.

Fig. 100. *Graphorkis lurida.* Gallery forest, Zutung, Kaduna St. February. 2.3x.

Fig. 101. *Podangis dactyloceras.* Gallery forest, Zutung, Kaduna St. January. 1x.

Fig. 102. *Podangis dactyloceras.* 1.75x.

Lateral sepals: Broadly and obliquely ovate-acute, apex sometimes incurved, 5 mm long, 3 mm wide, papillose, with dense and fine hairs, deep purple brown with indistinct white or grey markings, base whitish or purplish cream, with dark purple patches. Back of all sepals pale purple with small purple spots.
Petals: Linear-acute, with dense and fine hairs, 5-11 mm long, 0.5-0.7 mm wide, deep purple.
Lip: Ovate-acuminate, apex much reflexed, 4 mm long, 1.5 mm wide. Length of apical part 1 mm long, 0.5 mm wide. Basal part of lip pectinate. Whole lip cream-coloured with purple spots and purple margins.
Column: Purple with white markings, winged.
Distribution: S.

GENUS GRAPHORKIS

This genus is represented by a single species, *Graphorkis lurida,* in Africa. There are more species in Madagascar. It is a very common epiphyte in southern Nigeria, perhaps for the reason that it seems, to some extent, to have adapted to bush-burning. The pseudobulbs are quite large, conical, and have a yellowish colour. They are 6-9 cm high and 2-5 cm wide. As the plant is rather Eulophia-like, the genus was formerly known as 'Eulophiopsis'.

Graphorkis lurida (Sw.) O.Ktze. [Fig. 100]
Inflorescence branched, many-flowered, 20-50 cm long, appearing before the leaves in December-January. The flowers are about 1 cm in diameter. Easily cultivated.
Leaves: Lanceolate-acute, thin and stiff, ribbed, 200-450 mm long, 20-50 mm wide.
Sepals: Narrowly obovate-obtuse, 6 mm long, 2 mm wide, purplish brown, ventral side often greenish.
Petals: Obovate-obtuse, 6 mm long, 3 mm wide, brown. Ventral side often greenish mauve.
Lip: Trilobed, 6 mm long, 2.5 mm wide. Middle lobe retuse, yellow, with three keels at base. Side lobes erect, rounded, pale yellow or pale green.
Spur: Slightly flattened with truncate apex, 6 mm long, bent sharply forward, purple.
Column: Purple; operculum white.
Distribution: N + S.

GENUS PODANGIS

There is only one species in this genus, a small epiphyte with fleshy leaves that are laterally compressed and closely overlapping at the base. They rise in a tuft from a very short stem and spread fan-wise in a vertical plane. When not flowering, the plant is easily mistaken for a small *Rangaëris rhipsalisocia,* see p.87.

Fig. 103. *Calyptrochilum christyanum* var. *grandiflorum*. Forest, Oguta, Imo St. March. 1x.

Fig. 104. *Calyptrochilum christyanum* var. *'grandiflorum'*. 1.5x.

Fig. 105. *Calyptrochilum christyanum* var. *'viridum'*. In forest, Mgbidichi, Imo State. March. 0.8x.

Fig. 106. *Calyptrochilum christyanum* var. *'viridum'*. 2.5x.

Podangis dactyloceras (Rchb.f.) Schltr. [Plate 2.2; Fig. 101 & 102]
Inflorescences shorter than leaves, peduncle 1.5-3 cm long, 5-15-flowered. Pedicels 2-2.5 cm long (including ovary), white. The margins of all tepals, including the lip, are touching and form an ellipsoid structure. All tepals are white and translucent.
Leaves: Linear-acute, often falcate, 40-150 mm long, 5-12 mm wide.
Bracts: Small, triangular, scarious.
Dorsal sepal: Elliptical-obtuse, 4-5 mm long, 2-3 mm wide.
Lateral sepals: Elliptical-subacute, 4-6 mm long, 3-4 mm wide.
Petals: Obcordate-obtuse, often apiculate, 4-6 mm long, 3.5-5.0 mm wide (at apex).
Lip: Orbicular-obtuse, much concave, 5-6 mm long.
Spur: Tapering from a wide mouth, 8-14 mm long, apex much inflated and often 3- or 4-lobed, white.
Column: White; operculum pale green to yellow.
Distribution: N.

GENUS CALYPTROCHILUM

The two Nigerian representatives of this genus have long hanging stems and fleshy leaves with more or less parallel edges. The spur is swollen at the apex and the apical part is sharply deflexed. *Calyptrochilum christyanum* is quite common in both southern and northern Nigeria, and easily cultivated. It is difficult to display attractively though, due to its straggling habit and to the circumstance that the inflorescences appear under the leaves.

Calyptrochilum christyanum (Rchb.f.) Summerh.
A hanging epiphyte with long stems, often 1 m long or more, and numerous thick aerial roots. The leaves are very thick and leathery, unequally bilobed at the apex, with a depressed middle vein. The leaves are twisted at the base, so that the leaf blades are more or less in one plane. The trilobed lip is quadrangular-apiculate and has a short spur, about 1 cm long, with much inflated and abruptly deflexed apex, forming an elliptical sac. The author has found two varieties of this very common orchid, both widely distributed.

var. *grandiflorum* (Segerbäck 1165) [Fig. 103 & 104]
Leaves: Ligulate, nearly linear; lobes obtuse, sometimes (especially shorter lobe) truncate; 50-100 mm long, 15-28 mm wide, yellowish green or olive green; at an acute angle with stem.
Rhachis: Zig-zag-shaped, 20-40 mm long, 5-11-flowered, flowers rather persistent, faintly scented.
Dorsal sepal: Ovate (lanceolate)-apiculate, 9-10 mm long, 3-4 mm wide, pale yellow to pinkish white, spreading.
Lateral sepals: Obliquely ovate (lanceolate)-apiculate, 10 mm long, 4 mm wide, pale yellow to pinkish white, spreading.
Petals: Oblanceolate (oblong)-apiculate, 8 mm long, 3 mm wide, white.
Lip: Trilobed, 10-12 mm long, 6-7 mm wide, white, with yellow base. Middle lobe 7 x 7 mm, with two inflexed, obtuse front lobes having

Fig. 107. *Calyptrochilum emarginatum*. In forest, Oguta, Imo St.

Fig. 108. *Angraecum distichum*. Owerri, Imo St. November. 5.4x.

Plate 4
Above left: *Œceoclades saundersiana* (Lindl.) Garay & Taylor. 2x.
Above centre: *Bulbophyllum congolanum* Schltr. 2.7x.
Above right: *Nervilia umbrosa* (Reichb.f.) Schltr. 3x.
Below left: *Disperis johnstonii* Reichb.f. ex Rolfe. 5x.
Below right: *Diaphananthe plehniana* (Schltr.) Schltr. 5x.

PLATE 4

irregular margins and short deflexed point between the lobes. Side lobes rounded, erect, entire, 5 x 5 mm.
Spur: Apex green, rest of spur yellow.
Operculum: Pale yellow or white.
Distribution: N + S.

var. *viridum* (Segerbäck 1218) [Fig. 105 & 106]
Leaves: Subelliptic to narrowly ovate, convex, often with small point, edges of leaves recurved, 30-60 mm long, 15-20 mm wide, light green, at right angle with stem.
Rhachis: Indistinctly zig-zag-shaped, 6-9-flowered, 20-25 mm long, not scented, flowers shortly persistent.
Dorsal sepal: Ovate-acute, 5 mm long, 2.5 mm wide, greenish white, not much spreading.
Lateral sepals: Obliquely ovate-acuminate, 7 mm long, 3 mm wide, greenish white, not much spreading.
Petals: Oblong-lanceolate-apiculate, 5 mm long, 2 mm wide, greenish white, not much spreading.
Lip: Trilobed, white with green base. Middle lobe with two obtuse front lobes having entire margins and short, deflexed point between the lobes, 5 x 5 mm. Side lobes rounded, erect, entire, 2.5 mm long, 3 mm wide, white. Spur green.
Operculum: Dark brown.
Distribution: S.

Calyptrochilum emarginatum (Sw.) Schltr. [Fig. 107]
An epiphyte with long, hanging stems. The inflorescences are short, 3-5 cm long, they appear above the leaves and are densely many-flowered. The flowers are very fragrant, especially at night.
Leaves: Linear-lanceolate, hanging, slightly arching, very distinctly and unequally bilobed at apex, 100-170 mm long, 30-50 mm wide, thick and leathery.
Bracts: Closely overlapping, 4-7 mm long, dark brown.
Sepals: Broadly lanceolate-acute, 10 mm long, 3-5 mm wide, white.
Petals: Broadly oblanceolate-acute, 10 mm long, 3-5 mm wide, white.
Lip: Indistinctly trilobed in apical part, 10 mm long. Middle lobe short, pointed, side lobes rounded, overlapping. Whole lip laterally compressed, boat-shaped, yellow.
Spur: Much bent, slightly swollen at apex, 10 mm long.
Distribution: S.

GENUS ANGRAECUM

This is another large genus with more than 200 species, best represented in Madagascar. In Nigeria, there are some 20 species. The flowers in this genus are often quite large and scented. The lip is deeply concave ('cymbiform' = boat-shaped), and generally drawn out to a sharp point. The base of the lip usually envelops the column and at the rear end the lip gradually passes into a wide-mouthed spur. The column is deeply cleft into two lobes.

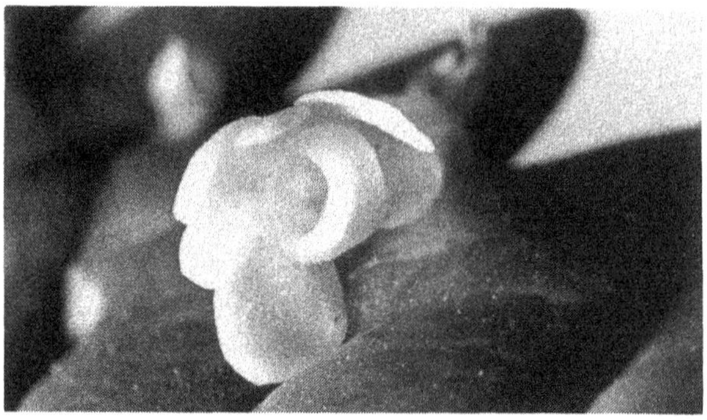

Fig. 109. *Angraecum distichum.* Roadside, Oguta, Imo St. August-November. 9.5x.

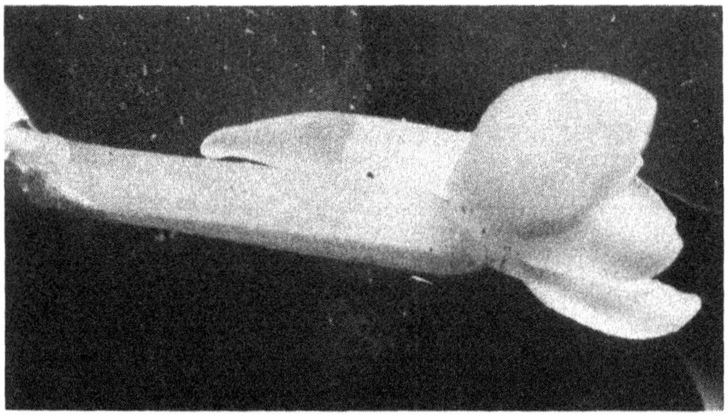

Fig. 110. *Angraecum distichum.* 10x.

Fig. 111. *Angraecum distichum* var. *'grandiflorum'.* On *Chlorophora excelsa,* in forest, Obinze, Imo St. February-April. 4.5x.

Fig. 112. *Angraecum distichum* var. *'grandiflorum'.* 4.75x.

Angraecum distichum Lindl. [Fig. 108 & 109 & 110]

Epiphyte with hanging habit, much branched. Branches arching, 10-30 cm long, often with leaves over entire length. Leaves opposite, very fleshy, laterally flattened, each successive leaf in flowering part of stem 0.1-0.2 cm above next one on the same side. Opposite leaves slightly overlapping at base. Groove on upper margin extending to within 0.1 cm of apex. Flowers solitary, numerous, small, faintly scented. The plant flowers often twice a year.

Leaves: Obliquely elliptical, 5-11 mm long, 3-7 mm wide, 1-1.5 mm thick.

Dorsal sepal: Elliptical-acute, slightly incurved, 2 mm long, 1.5 mm wide, white.

Lateral sepals: Elliptical-acute, slightly incurved, 3 mm long, 1.5 mm wide, white.

Petals: Elliptical-subacute, 2.5-3.0 mm long, 1 mm wide, white, slightly incurved.

Lip: Triangular-acute, cymbiform, indistinctly trilobed, 2 x 2 mm. Middle lobe pointed, with slightly recurved apex. Side lobes rounded, incurved. Whole lip white.

Spur: Tubular with tapering apex, very slightly inflated in the middle part, 2.5-3.0 mm long (about half to three-quarters of length of ovary).

Operculum: With pale green spot.

Distribution: S (+ N).

Angraecum distichum var. *grandiflorum* (Segerbäck 1191) [Fig. 111 & 112]

A smaller variety of the preceding species, but with larger flowers; found by the author in Imo State. The shorter branches are often erect, with a few flowers (1-4) at the ends of the branches. The flowers are about 1 cm wide. Each successive leaf is in contact with the next one on the same side of the branch.

Leaves: Obliquely and broadly lanceolate-obtuse, 6-7 mm long, 2-3 mm wide.

Dorsal sepal: Broadly lanceolate-obtuse, 5 mm long, 2 mm wide.

Lateral sepals: Obliquely oblong-lanceolate-acute, 5 mm long, 2 mm wide.

Petals: Oblong-obtuse, 4 mm long, 1-1.3 mm wide. All tepals white, much spreading, forming 90° angle with ovary.

Lip: Indistinctly trilobed, cymbiform, 4 mm long, middle lobe triangular-acute, apex recurved; side lobes rounded, incurved.

Spur: Tubular with tapering apex, longer than ovary, 7-10 mm long, apex often concealed by bract.

Distribution: S.

Angraecum aporoides Summerh.

This species is very similar to *Angraecum distichum* Lindl., but larger: the leaves are 1.5-2.5 cm long. Distribution: S.

Angraecum birrimense Rolfe. [Fig. 113]

A large, attractive epiphyte, often seen in southern Nigeria. The stem is long, often 2 m or more, slightly flattened and two-winged, generally hanging, with numerous aerial roots. The large fragrant flowers are usually two

Fig. 113 (above left). *Angraecum birrimense*. Mkpụrụ Forest, Imo St. May. 0.6x.
Fig. 114 (above right). *Angraecum subulatum*. 0.7x.
Fig. 115 (below left). *Angraecum subulatum*. Ukogugu Forest, Imo St. April. 7.6x.
Fig. 116 (below right). *Angraecum pyriforme*. Riverine Forest, Obinze, Imo St. April-May. 0.8x.

on each peduncle (4 cm long). In each pair of flowers, one opens much earlier than the other.

Leaves: Oblong-(elliptical)-lanceolate, unequally and obtusely bilobed at apex, 50-150 mm long, 20-45 mm wide, yellowish green, thickened. Middle vein depressed, forming a keel on lower surface of leaf.

Tepals: Lanceolate-acute, often with reflexed margins, 35-60 mm long, 3-9 mm wide, yellowish green to pale mauve.

Lip: Quadrate-(orbicular)-acuminate, 40-52 mm long, 30-38 mm wide, apical part 9-15 mm long. White, with yellowish-green base. Hypochilium with narrow green keel, continuing into spur.

Spur: Gradually tapering from a wide opening to the middle of its length, where its width is abruptly reduced to 2 mm; total length of spur 35-47 mm, pale green.

Distribution: S.

Angraecum subulatum Lindl. [Fig. 114 & 115]

An epiphyte with pendulous habit. Stems very slender, numerous, 30-60 cm long. The small, white flowers appear singly or in pairs.

Leaves: Subulate-terete, with a very sharp point, 50-130 mm long, 1-2.5 mm wide.

Sepals: Oblanceolate-acute, 4-5 mm long, 2 mm wide.

Petals: Linear-acute, 4 mm long, 0.5 mm wide.

Lip: Cymbiform, apiculate.

Spur: Slightly inflated in middle part, 15 mm long, 1 mm wide.

Distribution: S.

Angraecum pyriforme Summerh. [Plate 5.2; Fig. 116]

Epiphyte with long, hanging stems. Flowers always two on each peduncle (2 cm long). Of the two flowers on the same peduncle, one opens 10-14 days before the other. Flowers not resupinate. Tepals yellowish (or greenish) white, often with mauve apices. Older flowers pale yellow, with waxy appearance. Faintly fragrant at night.

Leaves: Narrowly lanceolate, unequally and obtusely bilobed at apex, 50-110 mm long, 10-22 mm wide, leathery, not much thickened.

Dorsal sepal: Lanceolate-acute, 10-13 mm long, 4 mm wide.

Lateral sepals: Obliquely ovate-acute, 9 mm long, 5 mm wide, margins often touching when flower opens, forming an orbicular-acute plane.

Petals: Obliquely ovate-apiculate, 10 mm long, 5 mm wide.

Lip: Ovate-acute, cymbiform, with longitudinal keel, 6-8 mm long, 2 mm wide.

Spur: Constricted in middle part, apex inflated, club-shaped, 10-14 mm long.

Operculum: Yellowish white.

Distribution: S.

GENUS MICROCOELIA

This is one of a few apparently quite leafless genera of orchids in which the numerous, densely matted roots have taken over the role of the leaves. On close inspection, however, rudiments of leaves can be seen in the shape of minute scales on the short stem. Five species (of a total of 27) have

Fig. 117 (above left). *Microcoelia caespitosa.* Uhoko Forest, Imo St. May-July. 3.8x.
Fig. 118 (above right). *Microcoelia caespitosa.* 1.4x.
Fig. 119 (below left). *Microcoelia guyoniana.* Gallery Forest, Assob Hills, Plateau St. March. 3x.
Fig. 120 (below right). *Diaphananthe curvata.* Forest, Naze, Owerri, Imo St. June. 0.7x.

been recorded for Nigeria. When not flowering, these little epiphytes are very easily overlooked.

Microcoelia caespitosa (Rolfe) Summerh. [Fig. 117 & 118]
Stem short, 1-3 cm long, with numerous roots and 1-10 inflorescences. The roots have short, longitudinal lines or spots, most clearly seen when the velamen is wet. The rhachis is 1-2 cm long and 5-10-flowered. The tepals are touching, forming a hood.
Dorsal sepal: Ovate-obtuse, 3 mm long, 2 mm wide, white, with well-defined green, central longitudinal line.
Lateral sepals: Ovate-obtuse (or acute), 4 mm long, 2.5 mm wide, white, with green line as in dorsal sepal.
Petals: Obovate-obtuse, 3 mm long, 2 mm wide, pure white.
Lip: Ovate-obtuse (or subacute), 3 mm long, 1.5 mm wide, white, with triangular green mark in apical part.
Spur: Constricted near mouth, much swollen in basal third, tapering towards apex which is again slightly inflated, 8-13 mm long, white.
Column: Green, rostellum white, operculum green.
Distribution: S.

Microcoelia guyoniana (Rchb.f.) Summerh. [Fig. 119]
Stem short, 2-6 cm long, with numerous roots. Inflorescences 2-13 cm long, rather densely many-flowered.
Bracts: Lanceolate-acute, clasping stem, 1 mm long, 0.5 mm wide, light brown.
Tepals: Ovate-obtuse to subacute, 2-4 mm long, 0.8-1.0 mm wide, white.
Lip: Oblanceolate-acute, apex deflexed, 2.5 mm long, white with green base.
Spur: Straight or slightly curved, tapering, apex obtuse but not swollen, 1-2 mm long, 0.2 mm wide, white, apex greenish or mauve.
Operculum: Bright brownish yellow.
Distribution: N.

GENUS DIAPHANANTHE

This is a genus with about 40 species. The Nigerian representatives are found mainly in the south. They often have translucent flowers, as the generic name indicates. The lip is generally broader than long, and provided with an irregular edge.

Diaphananthe curvata (Rolfe) Summerh. [Fig. 120]
Epiphyte in lower storey of rain forests. Stem very short, 0.5-1.5 cm long. Inflorescences 5-40 cm long, 5-25-flowered. All tepals green. Seed capsules 2-2.5 cm long, 0.3 cm wide. When not flowering, this species is easily mistaken for *Eurychone rothschildiana* or *Aërangis kotschyana,* the leaves of which are very similar.
Leaves: Oblanceolate (oblong), unequally and acutely bilobed at apex, with depressed middle vein, 40-170 mm long, 10-50 mm wide, dark green.
Dorsal sepal: Ovate-subacute, 4.5 mm long, 3 mm wide.
Lateral sepals: Ovate-subacute, 4 mm long, 2 mm wide.

Fig. 121. *Diaphananthe plehniana.* In forest, Ohosu, Bendel St. May. 3x.

Fig. 123. *Diaphananthe bidens.* 5x.

Fig. 122. *Diaphananthe bidens.* In forest, Oguta, Imo St. April. 0.6x.

Petals: Orbicular-(ovate)-subacute or obtuse, 3 mm long, 2 mm wide.
Lip: Suborbicular from a broad base, 2-3 mm long, 2.5 mm wide, with a rim-like callus at mouth of spur.
Spur: Tapering, 10-16 mm long.
Distribution: S.

Diaphananthe plehniana (Schltr.) Schltr. [Plate 4.5; Fig. 121]
Epiphyte with stout, winding stem, 1-5 cm long. Inflorescences 3-13 cm long, many-flowered. All tepals apricot-coloured, thin and translucent. Apex of rostellum violet.
Leaves: Narrowly oblanceolate to linear, indistinctly bilobed at apex (acutely or obtusely), thickened, sometimes falcate, 80-200 mm long, 8-35 mm wide, yellowish green.
Dorsal sepal: Ovate-apiculate, 6 mm long, 3 mm wide.
Lateral sepals: Lanceolate-apiculate, 7 mm long, 2.5 mm wide.
Lip: Quadrate, margins lacerate, 5-6 mm long, 4-8 mm wide.
Spur: Slightly swollen in apical part, 5-8 mm long.
Distribution: S.

Diaphananthe bidens (Sw.) Schltr. [Fig. 122 & 123]
Epiphyte with elongated, hanging stem, often nearly 1 m long, and numerous aerial roots. Inflorescences often numerous, 5-18 cm long, 15-30-flowered, pendulous. Flowers salmon-coloured, translucent, faintly and unpleasantly scented (like *Castanea sativa*).
Leaves: Oblong-lanceolate, distinctly unequally and acutely bilobed at apex, with prominent veins, 50-140 mm long, 20-45 mm wide.
Sepals: Lanceolate-acute, 3-7 mm long, 2 mm wide.
Petals: Narrowly lanceolate-acute, 6 mm long.
Lip: Quadrate, 4-7 mm long.
Spur: Slightly longer than lip, 7 mm long, 5.5 mm wide at base.
Distribution: S (common).

GENUS CHAMAEANGIS

This genus with five species in southern Nigeria includes orchids with rather short stems and long, fleshy narrow leaves, often bent so that their tips are nearly touching. The small flowers tend to appear in whorls on long spikes. The spur is inflated at the apex.

Chamaeangis vesicata (Lindl.) Schltr.
Epiphyte with numerous elongated inflorescences (up to 35 cm long), hanging, many-flowered. Flowers in pairs or groups of three in apical part of inflorescence, singly in basal part, yellow to orange, very fragrant. Flowers in August.
Leaves: Narrowly lanceolate-acute, much curved, 150-400 mm long, 7-30 mm wide, fleshy, V-shaped in section, olive green.
Sepals: Lanceolate-acute, 1.5-3.0 mm long.
Petals: Lanceolate-acute.
Lip: Lanceolate-obtuse, entire, 1.5-3.0 mm long.
Spur: 4.5-10 mm long, much swollen at apex, forming globular sac.
Distribution: S.

Fig. 124 (above left). *Chamaeangis* sp. nov. In forest, Ikom, Cross River St. May. 6.8x.
Fig. 125 (above right). *Chamaeangis* sp. nov. 9.4x.
Fig. 126 (below left). *Plectrelminthus caudatus.* 0.17x.
Fig. 127 (below right). *Plectrelminthus caudatus.* Mkpuru Forest, Naze, Imo St. May. 1.2x.

Chamaeangis sp.nov. ('nitida' Segerbäck 1169) [Fig. 124 & 125]
Short-stemmed (1-2 cm) epiphyte. Inflorescence very short, many-flowered. Rhachis slightly swollen. All parts of flowers greenish yellow.
Leaves: Narrowly lanceolate, slightly thickened, unequally and conniventlybilobed at subacute apex, 50 mm long.
Bracts: Triangular, 1 mm long, 2 mm wide, at base.
Dorsal sepal: Elliptical-acute (or apiculate), 2 mm long, 1.2 mm wide.
Lateral sepals: Elliptical-acuminate, 2 mm long, 1.5 mm wide.
Petals: Spatulate-obtuse, 1.5 mm long, 0.5 mm wide.
Lip: Triangular-cymbiform, apex deflexed, 3 mm long, 1 mm wide.
Spur: Much inflated, forming globular sac (2 mm ∅) at apex, 3.5 mm long.
Distribution: only found at Ikom, Cross River State.

GENUS PLECTRELMINTHUS

A genus with only one species in West Africa, *Plectrelminthus caudatus,* which is not uncommon. It is a large, sturdy epiphyte with a long, rather sparsely-flowered inflorescence and thick aerial roots. The flowers are quite large and conspicuous mainly because of the long, characteristically looped spur. The lip has a long, pointed middle lobe. The plant is easily cultivated in humid conditions.

Plectrelminthus caudatus (Lindl.) Summerh. [Fig. 126 & 127]
Rather short-stemmed epiphyte (stem 2-5 cm, rarely up to 15 cm long), with thick aerial roots (diameter 0.6-0.9 cm). Velamen light grey. Inflorescence zig-zag-shaped, 25-60 cm long, 4-10-flowered.
Leaves: Oblong-linear, obtusely and unequally bilobed at apex, imbricate at base, spreading fan-wise, 100-300 mm long, 15-40 mm wide.
Dorsal sepal: Narrowly lanceolate-acute, often acuminate, 30-50 mm long, 5-7 mm wide, green, with recurved margins.
Lateral sepals: Lanceolate-acute, often acuminate, 35-50 mm long, 4-5 mm wide, green, with recurved margins.
Petals: Narrowly lanceolate-acute, 30-35 mm long, 4-5 mm wide, green.
Lip: Trilobed, 40-45 mm long, white, obcordate, width across lamina 20-23 mm. Middle lobe acuminate, 11-12 mm long. Base of lip with two small, tooth-like processes. Side lobes rounded, lacerate.
Spur: 160-250 mm long, curved or spirally twisted, 2 mm wide, brownish-green.
Column: Elongated, 13-15 mm long, olive green with minute dark bristles on base. Rostellum 10 mm long, 2.5 mm wide, pale green with two pale brown keels in centre, apex pale brown. Pollinia two, with common stipes and viscidium. Viscidium small, globular, yellow. Length of stipes 8 mm. Operculum olive green, papillose, with triangular-acute, brown process.
Distribution: N + S.

GENUS AËRANGIS

This genus of handsome epiphytes has long, hanging and many-flowered inflorescences. The large, fragrant white flowers are star-shaped: all the

Fig. 128. *Aërangis kotschyana.* Gallery forest, Kajuru, Kaduna St. June. Seed capsules. 0.2x.

Fig. 130. *Rangaëris rhipsalisocia.* Gallery forest, Zutung, Kaduna St. January. 0.8x.

Fig. 129. *Rangaëris muscicola.* In forest, Ore, Ondo St. July. 3.2x.

Fig. 131. *Rangaëris rhipsalisocia.* On *Pentaclethra* sp., Avu, Imo St. January. 0.4x.

sepals and petals, including the flat, unlobed lip, are similar to each other. The spur is often very long and twisted. Of the existing 70 species, two have been recorded for Nigeria. Species of *Aërangis* are fairly easy to cultivate under humid, shady conditions.

Aërangis kotschyana (Rchb.f.) Schltr. [Fig. 128]
Stem 1-5 cm long. The leaves arise in a more or less fan-like tuft. The roots are thin and very long, often 1 m or more. Inflorescences 1-3, 18-45 cm long, 9-20-flowered. Flowers white, sometimes tinged pink, fragrant. This species is very similar to *Eurychone rothschildiana* and *Diaphananthe curvata,* when not flowering.
Leaves: Obovate to oblanceolate, unequally and subacutely (or obtusely) bilobed at apex, 80-200 mm long, 25-70 mm wide, greyish green to dark green, leathery and rather stiff, with depressed middle vein, forming keel on lower surface.
Tepals: Oblanceolate-acute, 20-30 mm long, reflexed.
Lip: Widened in middle part, apiculate, 25-30 mm long, 18 mm wide, reflexed.
Spur: Tubular, spirally twisted, slightly inflated in apical third, 130-250 mm long, pink to light brown.
Seed capsules: Long, slender.
Distribution: N.

Aërangis biloba (Lindl.) Schltr.
A widely distributed epiphyte found in northern as well as in southern Nigeria. The characteristic leaves are oblanceolate with two broadly rounded lobes at the apex. The inflorescences are quite long in this species as well, but the spur is much shorter, about 5 cm long. Flowering time: July-August.

GENUS RANGAËRIS

This genus has six species in southern Nigeria. It has rather large, fragrant white flowers in long inflorescences. The spurs are long and slender, and straight or only slightly curved, not twisted. The lip is similar to the other perianth segments, but wider. The genus does well in cultivation.

Rangaëris muscicola (Rchb.f.) Summerh. [Fig. 129]
Short-stemmed epiphyte with leaves arising in a fan-like tuft. Inflorescence often arching, 4-8-flowered, fragrant. Flowers white. This species is very similar to *Ancistrorhynchus cephalotes,* p.91, when not flowering.
Leaves: Narrowly lanceolate, obtusely and unequally bilobed at apex, 60-250 mm long, 10-25 mm wide, thin and stiff, V-shaped in section.
Dorsal sepal: Lanceolate-acute, 6 mm long, 3 mm wide.
Lateral sepals: Lanceolate-acute, 8 mm long, 3.5 mm wide.
Petals: Lanceolate-acute, 6 mm long, 2.5 mm wide.
Lip: Triangular-subacute, apex recurved, 7 mm long, 6 mm wide.
Spur: Slender, 50-90 mm long.
Distribution: S.

Fig. 132. *Cyrtorchis arcuata* ssp. *variabilis*. Anara F.R., Kaduna St. May.

Fig. 133. *C.arcuata* ssp. *variabilis*. 0.8x.

Fig. 134. *Cyrtorchis aschersonii*. Ukogugu Forest, Imo St. May. 0.6x.

Rangaëris rhipsalisocia (Rchb.f.) Summerh. [Fig. 130 & 131]
Short-stemmed epiphyte with thickened, much flattened leaves, closely overlapping at the base and forming a regular fan. When not flowering, this plant (especially the small, northern form) is very similar to *Podangis dactyloceras,* p. 69. Total height of plant 6-25 cm. Inflorescences 4-19 cm long, 5-18-flowered, very fragrant at night (clove-like scent). Flowers white.

Leaves: Linear-lanceolate-acute, falcate, 60-190 mm long, 5-15 mm wide.
Rhachis: Glabrous, intermittently canaliculate, with pedicels in grooves.
Pedicels: Sparsely pubescent, 15 mm long, including ovary.
Bracts: Triangular-obtuse, tightly clasping pedicel, scarious, 3-6 mm long, 2-5.5 mm wide (when flattened).
Sepals: Lanceolate-subacute, 8-12 mm long, 3-4 mm wide in the middle, 2 mm wide at the base.
Petals: Lanceolate-acute, 6-7 mm long, 2.5-3.0 mm wide.
Lip: Nearly hastate, 7-8 mm long, 4 mm wide at base, apical part drawn out to obtuse, thickened point, 1 mm wide, recurved.
Spur: 12 mm long, with acute, pale green apex, bent nearly 90° forward.
Operculum: White.
Distribution: N + S.

GENUS CYRTORCHIS

In this exclusively African genus of sturdy epiphytes with fleshy leaves, six species have been recorded for Nigeria (there are some 15 species in all). They do well in cultivation as they are not very demanding species, and the long sprays of white fragrant flowers are quite attractive. The flowers are symmetrical, the lip is very similar to the other perianth-segments, only slightly broader. The bracts are quite large, brown and triangular in shape.

Cyrtorchis arcuata ssp. *variabilis* Summerh. [Fig. 132 & 133]
A sturdy epiphyte with elongated stem (up to 50 cm, but generally much shorter), and numerous thick aerial roots. Inflorescences 1-4, 3-10 cm long, 3-12-flowered. Flowers white, turning orange, heavily scented at night. This species is very easy to cultivate.

Leaves: Linear-oblong, unequally and obtusely bilobed at apex, 50-250 mm long, 10-40 mm wide, yellowish green, thickened.
Bracts: Triangular-acute, up to 10 mm long, 19 mm wide, thin and dry, conspicuous.
Sepals: Lanceolate-acute, 20-31 mm long, 4-6 mm wide, apex more or less recurved.
Petals, including lip: Lanceolate-acute, 17-21 mm long, 5 mm wide, apex generally much recurved.
Spur: Slender, tapering, slightly curving, 30-50 mm long, pale green, apex pale mauve.
Distribution: N + S, common in N.

Cyrtorchis aschersonii (Krzl.) Schltr. [Fig. 134]
Epiphyte with elongated stem, 10-30 mm long. Inflorescences 5-14 cm

Fig. 135. *Eurychone rothschildiana.* Mkpụrụ Forest, Naze, Imo St. May. 1.5x.

Fig. 136. *Eurychone rothschildiana.* 1.1x.

PLATE 5

long, 5-15-flowered. Flowers white, fragrant at night, scent reminding of *Platanthera bifolia.*

Leaves: Very narrowly oblong, or linear, 80-220 mm long, 8-15 mm wide, very fleshy, more or less V-shaped in section, often unequally and obtusely bilobed at apex, sometimes with rolled-in margins; overlapping at base.

Bracts: Triangular, scarious, 10-15 mm long, 6-9 mm wide, dark brown.

Sepals: Lanceolate-acuminate, 12-16 mm long, 5-6 mm wide, much reflexed.

Petals: Lanceolate-acuminate, 11-12 mm long, 3-4 mm wide, much reflexed.

Lip: Lanceolate-acuminate, 12-15 mm long, 4-5 mm wide.

Spur: Tapering to a fine point from a broad base, slightly S-shaped, 15-40 mm long, pale green.

Distribution: S.

GENUS EURYCHONE

There is only one species of this small genus in Nigeria, *Eurychone rothschildiana.* This is a very handsome, large-flowered but rare epiphyte that grows in the lower storey of tropical rain forests. The dark green leaves are very similar to those of *Diaphananthe curvata* (p. 79) and *Aërangis kotschyana* (p. 85), which grow under the same conditions. Eurychone is not difficult to cultivate, provided the conditions are shady and humid. The delicate flowers are very fragrant, especially in the day-time (*Convallaria majalis*-like scent).

Eurychone rothschildiana (O'Brien) Schltr. [Fig. 135 & 136]

A short-stemmed (up to 7 cm) epiphyte with 1-3 inflorescences. Rhachis 3-7 cm long, 2-6-flowered.

Leaves: Obovate to oblanceolate, very unequally and subacutely bilobed at apex, 70-150 mm long, 50-70 mm wide, dark green, with wavy margins.

Dorsal sepal: Oblanceolate-acute, 23-25 mm long, 7 mm wide, white, with faint green longitudinal lines in centre.

Lateral sepals: Lanceolate-acute, 25 mm long, 7 mm wide, white.

Petals: Oblong-acute, 23-25 mm long, 7 mm wide, white, with faint green longitudinal lines in centre.

Lip: Ovate-emarginate, with irregular edge, 27 mm long, 20-25 mm wide; apex of lip white, middle part green, base dark brown.

Spur: Much constricted in middle, apex inflated, 20-25 mm long, 3 mm wide at apex, greenish brown.

Column: White, operculum very pale green.

Distribution: S.

Plate 5
Above left: *Tridactyle anthomaniaca* (Reichb.f.) Summerh. 4.6x.
Above centre: *Angræcum pyriforme* Summerh. 3x.
Above right: *Eulophia horsfallii* (Batem.) Summerh. 0.6x.
Below left: *Ansellia gigantea* var. *nilotica* (Baker) Summerh. 1.4x.
Below right: *Tridactyle tridactylites* (Rolfe) Schltr. 2.6x.

Fig. 137. *Ancistrorhynchus metteniae.* In forest, Owerri, Imo St. July. 3.6x.

Fig. 138. *Ancistrorhynchus capitatus.* Roadside, Oguta, Imo St. September. 8.5x.

Fig. 139. *Ancistrorhynchus cephalotes.* On *Pentaclethra* sp., Oguta, Imo St. April. 4.5x.

GENUS ANCISTRORHYNCHUS

In Nigeria, there are some eight species of this genus, which is well represented in the north. The flowers are typically found in dense heads at the base of the stem. The stiff brown bracts are large, and as long as the flowers. The rostellum is long, deeply split into two lobes, and has a typical J-bend.

Ancistrorhynchus metteniae (Kraenzl.) Summerh. [Fig. 137]
Epiphyte with short (4-10 cm) stem. Inflorescence short-stalked (0.6 cm long), many-flowered; flowers in very dense heads, with coconut-like scent.

Leaves: Lanceolate, unequally bilobed at apex, lobes entire or with a few, very fine teeth. Leaves 100-150 mm long, 15-25 mm wide, imbricate at base.
Bracts: Lanceolate-acute, longer than tepals, scarious, 10-13 mm long, 4-5 mm wide.
Sepals: Elliptical-obtuse, 5-6 mm long, 2.5-3 mm wide, white.
Petals: Elliptical to linear, truncate, 4 mm long, 2.5 mm wide, white.
Lip: Cymbiform, emarginate, very indistinctly trilobed, 4.5 mm long, 4 mm wide, white, with pale green transverse streak on mesochilium. Side lobes touching above column.
Spur: 4 mm long, with slightly inflated apex.
Distribution: N + S.

Ancistrorhynchus capitatus (Lindl.) Summerh. [Fig. 138]
Epiphyte with a short stem, up to 10 cm long. Inflorescence short-stalked with many small flowers in a dense head. All tepals white, with yellowish-green blotch on lip.

Leaves: Oblong-lanceolate, more or less equally bilobed at apex, with two acute teeth on each lobe, 140-200 mm long, 15-27 mm wide. Young leaves conduplicate.
Bracts: Ovoid-subacute, 13 mm long, 4 mm wide.
Dorsal sepal: Lanceolate-obtuse, 5 mm long, 2 mm wide.
Lateral sepals: Oblong-lanceolate-obtuse, 5 mm long, 1.5 mm wide.
Petals: Oblong-obtuse, 5 mm long, 1.5 mm wide.
Lip: Cymbiform, with indistinct, erect side lobes. Middle lobe triangular, fringed. Length of lip 6 mm, width 3 mm.
Spur: 8-11 mm long, much swollen at apex, forming an elliptical sac.
Distribution: S.

Ancistrorhynchus cephalotes (Rchb.f.) Summerh. [Fig. 139]
Short-stemmed epiphyte. Inflorescences with very short stalks and many small, densely tufted flowers. The plant is very similar to *Rangaëris muscicola,* p. 85, when not flowering. Flowers often fragrant, scent like that of Cymbopogon-leaves.

Leaves: Oblong-lanceolate to linear, obtusely and unequally bilobed at apex, thin but tough, V-shaped in section, 70-200 mm long, 8-19 mm wide.
Bracts: Scarious, shorter than tepals, 9 mm long.
Tepals: Oblong-obtuse, 7-8 mm long, 3 mm wide, white.

Fig. 141. *Ancistrorhynchus clandestinus.* In forest, Oguta, Imo St. June. 4x.

Fig. 140. *Ancistrorhynchus brunneo-maculatus.* Okahia Forest, Owerri, Imo St. May. 0.3x.

Fig. 142. *Ancistrorhynchus* sp. (*Segerb.* 1124). In forest, Benin, Bendel St. July. 2.5x.

Lip: Indistinctly trilobed, with wavy margin, 5-6 mm long, 5-7 mm wide, with green band across mesochilium. Side lobes touching above column.
Spur: 6-10 mm long, much swollen at apex, white.
Distribution: N + S.

Ancistrorhynchus brunneo-maculatus Rendle. (= *A. clandestinus* (Lindl.) Schltr.) [Fig. 140]
Epiphyte with a short stem, less than 5 cm long, often hanging, leaves arising in a fan-like tuft, closely overlapping. Inflorescences short, 2.0-2.5 cm long, many-flowered, arising from base of stem.
Leaves: Linear, very unequally and acutely bilobed at apex, 200-400 mm long, 7-8 mm wide, thick, slightly V-shaped in section.
Dorsal sepal: Ovate-acute, 3 mm long, 2 mm wide, pale brown, scurfy.
Lateral sepals: Lanceolate-acute, 4 mm long, 2 mm wide, pale brown, scurfy.
Petals: Oblong-acute, 3 mm long, 0.5 mm wide, pale brown with scurfy margins.
Lip: Oblong-acute, slightly constricted in middle, 4 mm long, 2 mm wide, pale brown, slightly scurfy.
Spur: S-shaped with slightly swollen apex, pale green, 5 mm long.
Distribution: S.

Ancistrorhynchus clandestinus (Lindl.) Schltr. [Fig. 141]
Epiphyte with pendulous habit. Stem short, less than 10 cm. Leaves in fan-like arrangement. Inflorescences from base of stem, 2.5-4.0 cm long, many-flowered.
Leaves: Very narrowly lanceolate and sharply pointed, 250-550 mm long, 15-23 mm wide, closely overlapping at the base, very unequally bilobed at apex (difference between lobes up to 30 mm), thick.
Sepals: Ovate-acute (or subacute), 4 mm long, 2.5 mm wide, white or pale green, scurfy, especially on back.
Petals: Lanceolate-obtuse (or subacute), 4 mm long, 1.5 mm wide, white, scurfy, especially on back.
Lip: Trilobed, 9 mm long, 4 mm wide, white, constricted in middle part, with green flecks at place of constriction. Middle lobe orbicular-obtuse with irregular margin. Basal end of middle lobe with upturned, erect margins. Side lobes rounded, erect.
Spur: 2-7 mm long, slightly bent, with incurved apex, scurfy.
Operculum: Very pale green, or yellow.
Distribution: N + S.

Ancistrorhynchus sp. (Segerbäck 1124) [Fig. 142]
Epiphyte with 4-10 cm long stem, leafy in upper part. Inflorescences short, with densely spaced flowers. All tepals white, lip with transverse green streak at place of constriction. Back of sepals and ovary scurfy.
Leaves: Very narrowly lanceolate, unequally and conniventsly bilobed at apex, with sharply pointed lobes; 75-150 mm long, 6-9 mm wide, closely imbricate at base, distinctly V-shaped in section, very thick.
Bracts: Lanceolate-acute, scarious, not exceeding length of ovary, 5-7 mm long, 1.5-2.5 mm wide.
Dorsal sepal: Lanceolate-obtuse, 3.5 mm long, 1.5 mm wide.
Lateral sepals: Lanceolate-acute, 4.5 mm long, 2 mm wide.

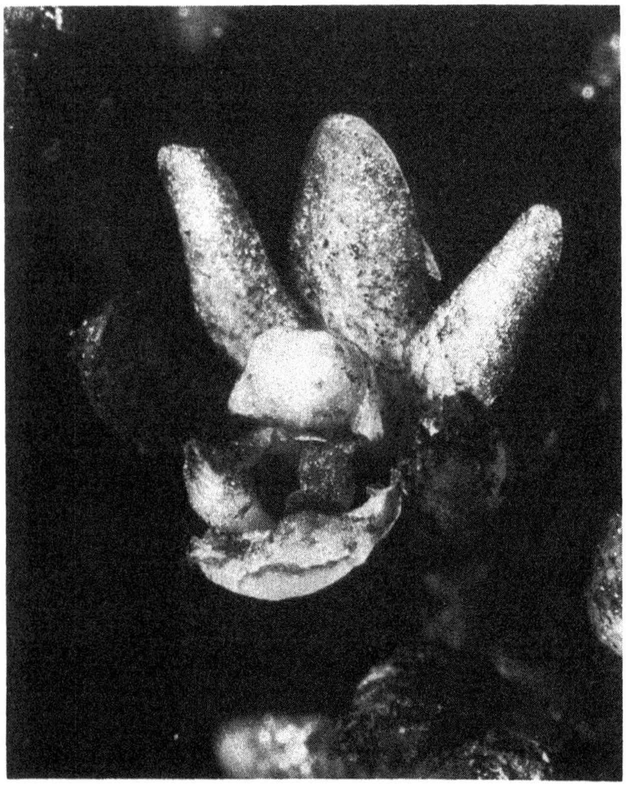

Fig. 143. *Ancistrorhynchus recurvus.* In forest, Oguta, Imo St. June. 11x.

Fig. 144. *Tridactyle tridentata.* Mkpụrụ Forest, Naze, Imo St. June. 3.75x.

Fig. 145. *Tridactyle tridactylites.* Gallery forest, Kufana, Kaduna St. April. 3.5x.

Petals: Linear-obtuse, 3 mm long, 0.7 mm wide, with recurved margins.
Lip: Trilobed, 6.5 mm long, 3.5 mm wide, thickened and constricted between middle and side lobes. Middle lobe rounded, concave, with irregular margins. Side lobes entire, obtuse, erect.
Spur: Reflexed, swollen at apex, 4 mm long, pale green, slightly scurfy.
Distribution: S.

Ancistrorhynchus recurvus Finet. [Fig. 143]
Epiphyte with pendulous habit. Stem short, up to 10 cm long. Inflorescences from base of stem, numerous, many-flowered, 2-3 cm long.
Leaves: Narrowly lanceolate, more or less unequally (up to 40 mm difference) and obtusely bilobed at apex, imbricate at base, 150-400 mm long, 25-30 mm wide, slightly thickened.
Sepals: Ovate-obtuse, 4 mm long, 2 mm wide, very pale green, scurfy, especially dorsal side.
Petals: Lanceolate-obtuse, 4 mm long, 1.5 mm wide, very pale green, scurfy, especially dorsal side.
Lip: Trilobed, 3-5 mm long, 2-4 mm wide. Side lobes erect, making contact with column. Middle lobe triangular. Lip constricted where middle lobe meets side lobes.
Spur: Short, bent, 2 mm long, pale green, slightly scurfy.
Distribution: S.

GENUS TRIDACTYLE

The plants in this genus have long, straggling stems which are often leafless except for the apical part. The lip is generally divided into three lobes, and provided with two 'auricles' at the base (see Fig. 4e). Ten species have been found in Nigeria. In all, there are about 35 species in this purely African genus.

Tridactyle tridentata (Harv.) Schltr. [Fig. 144]
Epiphyte with pendulous habit. Stems 1 m long or more, much branched. Inflorescences very short-stemmed, 1-3-flowered. All tepals pale yellowish-brown. Column sometimes orange.
Leaves: Subulate-terete, with narrow groove along upper surface, 90-150 mm long, 1-2 mm wide.
Dorsal sepal: Oblong-acute, 1.5-3.5 mm long, 1.5 mm wide.
Lateral sepals: Ovate-subacute, slightly falcate, 1.5-3.5 mm long, 2 mm wide.
Petals: Narrowly lanceolate-subacute, lacerate at apex, 3 mm long, 0.5 mm wide.
Lip: Trilobed, linear, 2.5-3.0 mm long, 1 mm wide, with two triangular auricles. Middle lobe triangular, with acute apex bent sharply upwards. Side lobes bent slightly upwards, very acute, pale green.
Spur: Very slender, 6-8 mm long.
Distribution: S.

Tridactyle tridactylites (Rolfe) Schltr. [Plate 5.5; Fig. 145]
Epiphyte with hanging, elongate, straggling stems, often branched, leafy

Fig. 146. *Tridactyle anthomaniaca.* Roadside, Owerri, Imo St. April. 0.9x.

Fig. 147. *Tridactyle crassifolia.* On *Chlorophora excelsa.* Ukogugu Forest, Imo St. June. 4.3x.

mainly at apex. Inflorescences numerous, 3-4 cm long, 6-12-flowered. Flowers pale mauve to orange, fragrant at night.
Leaves: Linear-ligulate, thin, folded lengthwise at base, narrowing at very unequally bilobed apex, 70-120 mm long, 8-10 mm wide.
Dorsal sepal: Lanceolate-subacute, 3-5 mm long, 1.5 mm wide.
Lateral sepals: Obovoid-apiculate, 3-5 mm long, 2 mm wide.
Petals: Linear–(sub)acute, 2.5-4.0 mm long, 0.8 mm wide.
Lip: Trilobed in apical part, side lobes divergent, longer than middle lobe. Length of lip 4 mm, width across mesochilium 2 mm; small auricles near opening of spur. Middle lobe triangular, 2 mm long, width at base 1.2 mm. Side lobes thin, tapering, length 2.2 mm, width at base 0.5 mm.
Spur: Thin, tapering, 7-20 mm long, 0.5 mm wide, light green.
Distribution: N.

Tridactyle anthomaniaca (Rchb.f.) Summerh. [Plate 5.1; Fig. 146]
Epiphyte with elongated, hanging stems, often branched. Inflorescence with very short peduncle, 0.1-0.2 cm long, 2-4-flowered. Flowers pale orange, yellowish green or pale green.
Leaves: Strap-shaped, unequally and obtusely bilobed at apex, thick and leathery with recurved edges, 20-80 mm long, 10-20 mm wide.
Dorsal sepal: Linear-acute, 4-7 mm long, 2.5 mm wide.
Lateral sepals: Lanceolate-acute, 4-6 mm long, 2 mm wide.
Petals: Linear-acute, 4-5 mm long, 2.5-3.0 mm wide.
Lip: Lanceolate-acute, with two distinct, acute basal auricles, 4-6 mm long, 2.5-3.0 mm wide.
Spur: Constricted in the middle, with inflated apex; 6-15 mm long.
Column: Orange or pale green.
Distribution: S.

Tridactyle crassifolia Summerh. [Fig. 147]
Epiphyte with elongated, hanging, straggling stems, often branched. Inflorescences very short-stemmed, 1-3-flowered.
Leaves: Strap-shaped, tapering towards unequally bilobed apex. Longer lobe obtuse, shorter lobe obtuse or truncate. Whole leaf 50-80 mm long, 8-10 mm wide, very fleshy, V-shaped in section.
Sepals: Triangular-subacute, 3-6 mm long, 2.5-3.0 mm wide, pale mauve to green.
Petals: Indistinctly spatulate-acute, 4-5 mm long, 1.3 mm wide, green.
Lip: Very indistinctly trilobed, lanceolate-subacute, 5 mm long, 2 mm wide, green, with small, ovate, obtuse to acute, basal auricles. Margin of middle lobe very finely fringed. Side lobes barely visible, very small.
Spur: Slender, very slightly inflated in middle, 10-16 mm long.
Column: 2.5 mm long, pale green to mauve, operculum brown. Rostellum curved, pale green.
Distribution: S.

CHAPTER FOUR

Botanical glossary

Acuminate: with a long, tapering point, see figure 148.
Acute: ending in a sharp point, see figure 148.
Adnate: grown together, fused with (said of different parts).
Adpressed: pressed close to.
Amplexicaul: clasping the stem.
Anterior: facing away from the axis. A. lobe of petal, see figure 11.
Anther: the part of the stamen that contains pollen.
Anther canal: antheric process.
Anther cap (= Operculum): anther modified into little lid, covering pollinia.
Antheric process: see figure 11.
Apex pl. Apices: the end farthest from the point of attachment.
Apiculate: ending in a small point, see figure 148.
Appendage: a subordinate or external part.
Assimilate: absorb and build up foodstuffs into complex constituents of an organism.
Auricle: an ear-like part.
Axil: the angle between the upper side of an organ, e.g. a leaf, and an axis, e.g. a branch.
Axillary: arising in the axil.
Basal: at the lower end.
Bearded: with long hairs.
Bifid: forked, split into two parts.
Bilobed: with two lobes, see figure 148.
Bract: modified leaf at the base of a flower or of a branch of an inflorescence.
Bracteole: small bract.
Bullate: (of leaf, petal, etc.) having inflated areas between the veins.
Callus pl. Calli: a hardened projection on the lip.
Calyx: outermost part of a flower, the sepals.
Canaliculate: with a longitudinal groove.
Capsule: dry fruit with many seeds in more than one section, splitting open when ripe (in orchids).
Carpel: female reproductive organ of flowering plants, consisting of ovary and stigma.
Caudate: with a slender, tail-like appendage, see figure 148.
Caudicle: stalk of pollinium formed from pollinial tissue.
Chlorophyll: green colouring matter of flowering plants; it enables the plants to build up carbohydrates from carbon dioxide and water, using energy from sunlight. This process is called 'photosynthesis'.
Claw: narrow, basal part of petal, especially of lip.
Column: union of stamens and styles in orchids.
Concave: hollow, with a surface that curves inward.
Conduplicate: folded together lengthwise.
Connivent: with the ends inclined towards each other.
Conspicuous: easily seen, attracting notice.
Convex: with a surface that curves outwards.
Cordate: heart-shaped and wider at the base, see figure 149.
Coriaceous: leathery.

Corolla: collective name for the petals in a flower.
Crenate: with rounded teeth.
Cuspidate: having a rigid, sharp point.
Deciduous: shedding leaves at a certain season or stage of growth.
Deflexed: bent downward.
Dentate: with pointed, outward directed teeth.
Distal: situated away from (point of attachment).
Distichous: placed opposite each other in two vertical rows.
Divergent: going in different directions.
Dorsal: of, on, or near the back.
Dorsal sepal: see figure 4.
Elliptical: see figure 149.
Elongate: long and narrow.
Emarginate: with a notch at the tip, see figure 148.
Embryo: the rudimentary plant in a seed.
Endosperm: nourishment for embryo, surrounding it in a seed.
Entire: without teeth or lobes.
Epichilium or Epichile: front part of lip.
Epiphyte: plant, using another plant for support without obtaining food from it.
Erect: directed upward.
Evolution: development of plants from earlier forms by means of inherited variations.
Falcate: curved like a sickle, see figure 149.
Family: a division in the realm of plants below 'order' and above 'genus', consisting in a number of similar genera. Family names generally have the ending '-aceae'.
Filiform: thread-like, very thin.
Fimbriate: fringed (with thicker processes than hairs).
Flabellate: fan-shaped.
Flora: the plants of a particular area, country, region, period, etc. Also a book treating such a selection of plants.
Form: smallest group used in plant classification: a category within a species with only small variations from the type, e.g. colour of petals.
Fungus pl. Fungi: simply organized plant lacking chlorophyll (e.g. mushroom, mould, yeast), made of cellular filaments called hyphae; reproduction is by means of spores.
Fused: joined together.
Genus pl. Genera: one of the groups used in classifying plants. Consists of a number of similar species. Similar genera are grouped in a family.
Germinate: to sprout, to start growing.
Glabrous: hairless.
Habitat: a particular environment inhabited by a plant.
Hastate: shaped like a halberd: triangular with spreading basal angles, see figure 149.
Hibernate: to pass the dry season in a dormant condition.
Horticulture: the science of cultivating fruits, vegetables and decorative plants.
Hyaline: thin and translucent, glassy.
Hybrid: result of crossing of different species (or genera).
Hyphæ: see Fungus.
Hypochilium or Hypochile: basal part of lip.
Imbricate: overlapping, like roof-tiles.
Incurved: curved inward.
Inflated: blown up.
Inflexed: bent inward.
Inflorescence: flowering shoot.
Keel: a sharp ridge.
Labellum: lip.
Lacerate: with an irregular edge, as if torn.

Lamella pl. Lamellæ: a thin plate.
Lanceolate: narrow and gradually tapering toward apex, see figure 149.
Latex: a milky juice produced by plants.
Lax: (of inflorescence) loose, open.
Ligulate: strap- or tongue-shaped, see figure 149.
Linear: long, narrow and with parallel sides, see figure 149.
Lip: the lower or middle petal of an orchid.
Lobe: a (rounded) projection or division.
Mesochilium or Mesochile: middle part of lip.
Monopodial: having single main axis continuing to grow at apex.
Motile: having an inherent power of motion.
Mycorrhiza: an association of a fungus with a root of a higher plant.
Nectar: a sugary liquid attractive to insects.
Nectarium or Nectary: part of flower secreting nectar.
Node: part of stem from where one or more leaves arise.
Obcordate: heart-shaped, with the broad end upward, see figure 149.
Oblanceolate: narrow and gradually widening toward apex, see figure 149.
Oblong: see figure 149.
Oblique: with unequal sides.
Obovate: egg-shaped, with the broader end upward, see figure 149.
Obtuse: blunt, see figure 148.
Operculum = Anther Cap.
Orbicular: round and flat, see figure 149.
Ovary: hollow lower part of carpel.
Ovate: egg-shaped, with the broader end at the base, see figure 149.
Panicle: an irregularly branched inflorescence.
Paniculate: having a panicle.
Papilla pl. Papillæ: a tiny protruding cell.
Papillose: covered with papillæ.
Parasite: a plant, growing on another plant, and obtaining food from it.
Pectinate: comb-like.
Pedicel: a stalk of an individual flower in an inflorescence.
Peduncle: a stalk of an inflorescence or of a solitary flower.
Perianth: sepals and petals.
Persistent: remaining attached for a long time.
Petal: see figure 4.
Petiole: a stalk of a leaf.
Photosynthesis: the formation of organic compounds from carbon dioxide and water in green plants, using energy from sunlight.
Plicate: folded like a fan.
Pollinate: to convey pollen from an anther to a stigma.
Pollinium pl. Pollinia: pollen grains clumped together.
Porrect: extended horizontally.
Posterior: near the axis.
Posterior lobe: (of petal), see figure 11.
Prostrate: lying flat (on the ground or on the substrate).
Pseudobulb: swollen part of stem (internode), often resembling a bulb.
Puberulent: very finely pubescent.
Pubescent: covered with soft, downy hairs.
Quadrate: square or nearly square.
Raceme: inflorescence with elongated axis, carrying simple pedicels.
Rachis = Rhachis, *q.v.*
Recurved: curved downward and/or backward.
Reflexed: bent downward and/or backward.
Resupinate: (of flower) having lip downward and dorsal sepal upward.

Reticulate: having a structure like the meshes of a net.
Retuse: with a rounded end, the centre of which is depressed.
Rhachis: the principal axis of an inflorescence.
Rhizome: rootlike, often thickened horizontal stem under the ground or along the substrate, with scale-like leaves; serving as a means of survival over several seasons.
Rostellum: projection of upper edge of stigma in front of anther.
Rugose: wrinkled.
Saprophyte: plant which feeds on organic matter in solution from dead or decaying tissue of other plants.
Scabrous: rough to the touch from being covered with very small stiff hairs or points.
Scale: a thin dry structure.
Scape: a leafless flowering stalk.
Scarious: thin and dry, withered.
Scurfy: covered with very small scales.
Secund: bearing flowers on one side only.
Sepal: see figure 4.
Sessile: without a stalk.
Setaceous: bristle-like.
Sheath: a leaf-base, tightly wrapped around the stem.
Spatulate: spoon-shaped (but often flat), see figure 149.
Species: a single distinct kind of plant.
Spur: extension backwards or downwards of petal (or sepal).
Stamen: the male reproductive organ in a flower.
Staminodium or Staminode: a sterile, rudimentary stamen.
Stelidium: vertical projection from apex of column.
Stem: the main axis of a plant.
Sterile: not capable of producing pollen.
Stigma: topmost part of ovary, receiving the pollen.
Stipes: stalk of pollinium.
Style: upper part of ovary terminating in stigma.
Sub-: prefix, meaning 'not quite', 'nearly', as in 'subacute'.
Subspecies: group within a species used in plant classification. The members of the group have certain characteristics in common, and differ from other members of the species.
Subulate: awl-shaped (slender and gradually tapering to a fine point), see figure 149.
Succulent: with fleshy appearance, having water-storing tissues.
Suppressed: undeveloped.
Sympodial: having a growth with yearly lateral shoots.
Taper: to decrease gradually in width.
Tepal: sepal or petal (generally excluding lip).
Terete: circular in cross section.
Terminal: at the end of a stem.
Terrestrial: growing in the ground.
Tessellate(d): checkered.
Translucent: letting light pass through.
Transverse: placed across.
Trilobed: with three lobes.
Truncate: appearing as if abruptly cut off at the tip, see figure 148.
Tuber: swollen part of underground stem, with dormant buds. Tubers serve as storing organs for nutrients.
Undulate: having a wavy margin.
Variety: group used in plant classification (below subspecies).
Velamen: water-absorbing outer cover of aerial root, consisting of several layers of dead cells.
Ventral: of, or belonging to, the inner face.

Verrucose: warty, wart-like.
Viscidium: part of rostellum forming a sticky disc, to which pollinia are attached.
Whorl: arrangement of leaves, etc. in a circle.
Wing: flattened extension of stem.

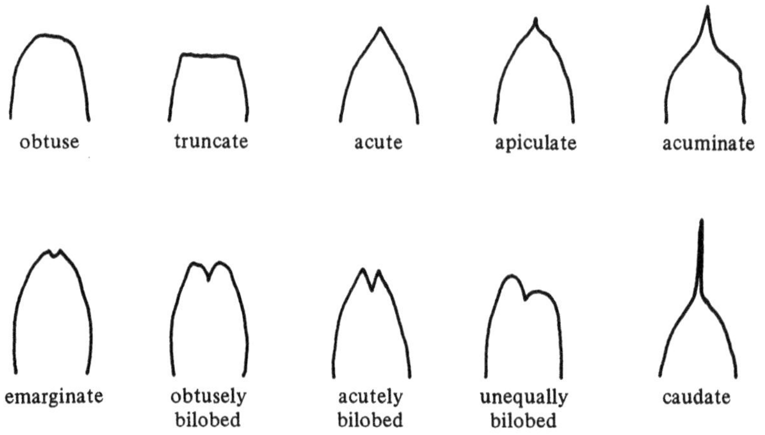

Fig. 148. Terms describing tips of leaves, perianth segments, etc.

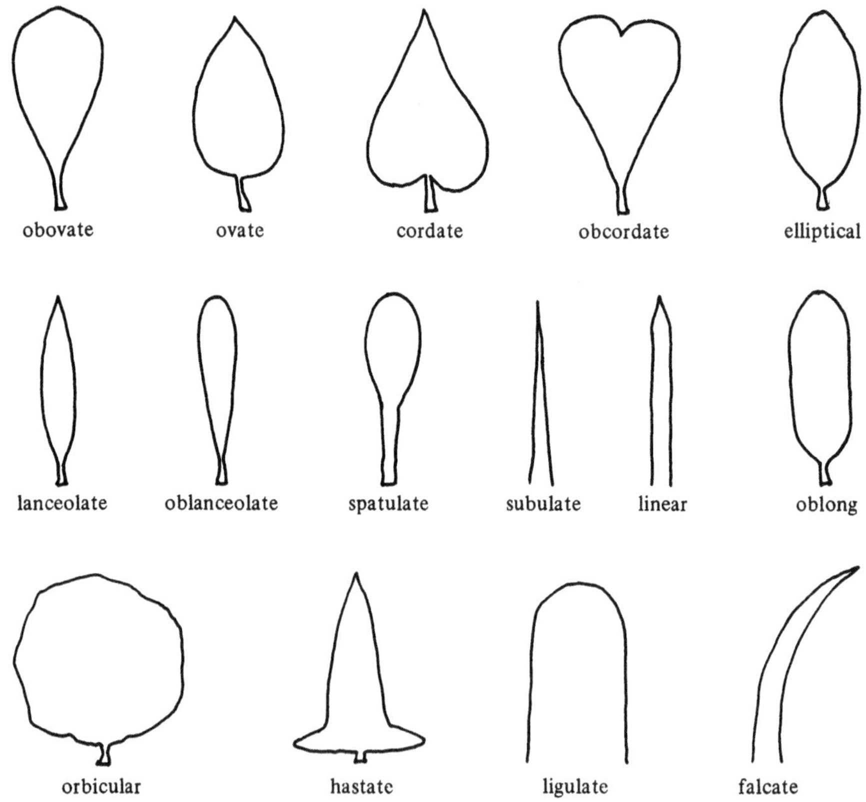

Fig. 149. Terms describing shapes of leaves, perianth segments, etc.

C.Z.TANG & P.J.CRIBB

Check-list of Nigerian orchids

Species	Distribution in Nigeria	Reference herbarium specimen (K)
Holothrix aphylla (Forsk.) Rchb.f.	N	King 102
Schwartzkopffia pumilio (Lindl.) Schltr.	N	King 110
Brachycorythis macrantha (Lindl.) Summerh.	N, S	Keay FHI 25960, Gillett 15312
B.tenuior Rchb.f.	N	Dalziel 835
B.conica (Summerh.) Summerh.	N	Chapman 2
B.sceptrum Schltr.	N, S	King 145, Sanford 1785/65
B.buchananii (Schltr.) Rolfe	N, S	Chapman 4, King 158
B.pleistophylla Rchb.f.	N	King 42
B.ovata subsp. *schweinfurthii* (Rchb.f.) Summerh.	N	Dalziel 561
B.pubescens Harv.	N	Dalziel 571
Habenaria filicornis Lindl.	N	Lely P330
H.chlorotica Rchb.f.	N	Dalziel 232
H.macrandra Lindl.	S	Mann 518
H.zambesina Rchb.f.	S	Keay FHI 28038
H.procera (Sw.) Lindl.	S	Keay FHI 22589
H.weilerana Schltr.	S	Talbot 844
H.gabonensis Rchb.f.	N	Chapman 22
H.dinklagei Kränzl.	S	Jones FHI 20721
H.nigerica Summerh.	N	Keay FHI 25993
H.peristyloides A.Rich.	N, S	Chapman 3, Stone 54
H.englerana Kränzl.	N, S	King 186, Westwood 268
H.prionocraspedon Summerh.	S	Rosevear 61/29
H.parva (Summerh.) Summerh.	N	Lely P337
H.mannii Hook.f.	N, S	Lely 506, Stone 82
H.longirostris Summerh.	N	Lely 462
H.barrina Ridl.	S	Punch s.n.
Habenaria papyracea Schltr.	N	King 100
H.buettnerana Kränzl.	N, S	King 96, Savory 31
H.malacophylla Rchb.f.	N, S	Killick 21, Talbot 3355
H.bongensium Rchb.f.	N	Lely 407
H.genuflexa Rendle	N, S	Lely 508, Savory 191
H.dalzielii Summerh.	N	King 92
H.ichneumonea (Sw.) Lindl.	N	King 175
H.pauper Summerh.	N	Chapman 1
H.huillensis Rchb.f.	N	Lely P732
H.linguiformis Summerh.	N	King 27
H.chirensis Rchb.f.	N	King 28
H.phylacocheira Summerh.	N	Lely 451

Species	Distribution in Nigeria	Reference herbarium specimen (K)
H.clavata (Lindl.) Rchb.f.	N	G.V.Summerhayes 134
H.holubii Rolfe	N	Dalziel 839
H.cirrhata (Lindl.) Rchb.f.	N	Lely P615
H.laurentii De Wild.	N, S	Lely 319, Onochie FHI 40879
H.cornuta Lindl.	N	Dalziel 228
H.walleri Rchb.f.	N	Lely 315
H.occidentalis (Lindl.) Summerh.	N	Barter 148
H.lelyi Summerh.	N	Lely 457
H.lecardii Kränzl.	N	Dalziel 566
H.keayi Kränzl.	N	King 30a
H.macrura Kränzl.	N	Dent Young s.n.
H.armatissima Rchb.f.	N	E.Vogel s.n.
Platycoryne paludosa (Lindl.) Rolfe	N	Barter 1479
P.megalorrhyncha Summerh.	N	Chapman 11
Disa equestris Rchb.f.	N	King 90
D.nigerica Rolfe	N	Lely 446
D.hircicornis Rchb.f.	N	King 161
D.erubescens Rendle	N	Chapman 12
D.welwitschii Rchb.f.	N	Dent Young s.n.
Satyrium carsonii Rolfe	N	Lely 344
S.volkensii Schltr.	N	Lely P318
S.crassicaule Rendle	N	Chapman 20
S.atherstonei Rchb.f.	N, S	Chapman 9, King 157
Disperis johnstonii Rchb.f.	N	Lely 485
D.anthoceros Rchb.f.	N	King 187
Vanilla africana Lindl.	S	Barter 47
V.ramosa Rolfe	S	Barter 20134
V.crenulata Rolfe	S	Miles s.n.
V.nigerica Rendle	S	Talbot 776
Nervilia petraea (Afzel. ex Pers.) Summerh.	N	King 81
N.reniformis Schltr.	N, S	King 111, Binuyo FHI 36921
N.fuerstenbergiana Schltr.	N	King 95
N.adolphii Schltr.	N	King 112b
N.kotschyi (Rchb.f.) Schltr.	N	King 107
N.umbrosa (Rchb.f.) Schltr.	N, S	Barter 1540, Keay FHI 23430
Auxopus kamerunensis Schltr.	N, S	Barter s.n., Sanford WS/49
A.macranthus Summerh.	S	Brenan 8535a
Manniella gustavii Rchb.f.	S	Holland 194
Platylepis glandulosa (Lindl.) Rchb.f.	S	Killick 246
Cheirostylis lepida (Rchb.f.) Rolfe	S	Talbot 870
Zeuxine elongata Rolfe	S	Talbot 1360
Z.africana Rchb.f.	N	Dalziel 447
Hetæria stammleri (Schltr.) Summerh.	S	Richards 3072
H.tetraptera (Rchb.f.) Summerh.	S	Ross 284
H.occidentalis Summerh.	N, S	Dalziel 231, Talbot 920
Corymborkis corymbosa Thou.	S	Onochie FHI 19106
Malaxis maclaudii (Finet) Summerh.	N, S	King 103, Sanford WS/ 1503/65

Species	Distribution in Nigeria	Reference herbarium specimen (K)
M.chevalieri Summerh.	S	Keay FHI 16229
M.prorepens (Kränzl.) Summerh.	N, S	King 177, Sanford 1601/65
M.katangensis Summerh.	N	King 98
Liparis tridens Kränzl.	S	Talbot 872
L.caillei Finet	N	King 86
L.platyglossa Schltr.	S	cult. sp.
L.nervosa (Thunb.) Lindl.	N, S	G.V.Summerhayes 158, Jones FHI 20722
L.epiphytica Schltr.	S	Lloyd 3
L.suborbicularis Summerh.	S	Talbot 871
Ansellia africana Lindl.	S	Talbot 1387
A.gigantea Rchb.f. var. *nilotica* (Bak.) Summerh.	N	Kennedy 2994
Polystachya paniculata (Sw.) Rolfe	S	Keay FHI 25371
P.albescens Ridl. subsp. *albescens*	S	Head 96
P.cooperi Summerh.	S	Cooper 2a
P.laxiflora Lindl.	S	Barter 2126
P.stricta Rolfe	N	Hepper 2882
P.bifida Lindl.	S	Cooper 84
P.rhodoptera Rchb.f.	S	Kennedy 1917
P.saccata (Finet) Rolfe	N	G.V.Summerhayes 90
P.elegans Rchb.f.	S	Keay FHI 25376
P.coriscensis Rchb.f.	S	Talbot s.n.
P.alpina Lindl.	S	Talbot 835
P.seticaulis Rendle	S	Talbot 926
P.steudneri Rchb.f.	N	King s.n.
P.adansoniae Rchb.f.	N, S	E.W.Jones 223, Talbot 867
P.polychaete Kränzl.	S	Talbot 927
P.ramulosa Lindl.	S	Talbot 3647
P.odorata Lindl. var. *odorata*	S, N	King 154, Barter 1483
P.odorata var. *trilepidis* Summerh.	S	Jones FHI 20729
P.tessellata Lindl.	N, S	King 54, Mann s.n.
P.mukandaensis De Wild.	S	Thomas 1902
P.modesta Rchb.f.	N, S	Cole 13, Keay FHI 22429
P.golungensis Rchb.f.	N, S	King 22, Keay 28294
P.dolichophylla Schltr.	N, S	Hamilton s.n., Latilo FHI 26118
P.subulata Finet	N	Keay FHI 22898
P.parva Summerh.	N	King 198
P.affinis Lindl.	S	Burtt B27
P.obanensis Rendle	S	Talbot 930
P.camaridioides Summerh.	S	Wright 101
P.supfiana Schltr.	S	Talbot s.n.
P.galeata (Sw.) Rchb.f.	S	Brenan 9116
Stolzia elaidum (Lindl.) Summerh.	S	Barter 73
S.repens (Rolfe) Summerh.	S	Sanford 145/66
Ancistrochilus thomsonianus (Rchb.f.) Rolfe	S	Talbot 88
A.rothschildianus O'Brien	S	Ross 227
Bulbophyllum flavidum Lindl.	S	King 184
B.schimperanum Kränzl.	S	Schlechter 15755
B.josephii (Kuntze) Summerh.	S	Horwood 165

Species	Distribution in Nigeria	Reference herbarium specimen (K)
B.calvum Summerh.	N	Wimbush in King 178
B.nigritianum Rendle	S	Talbot 933
B.buntingii Rendle	N, S	King 200, Talbot 935
B.pavimentatum Lindl.	S	Mann 519
B.winkleri Schltr.	N, S	Hamilton s.n., Winkler 157
B.porphyroglossum Kränzl.	S	Talbot 3288
B.intertextum Lindl.	N, S	King 58, Mann 527
B.tenuicaule Lindl.	N	Wimbush in King 150
B.cochleatum Lindl.	N, S	Hepper 1507, Talbot 931
B.distans Lindl.	S	Mann 525
B.calamarium Lindl.	S	Wright 137
B.phaeopogon Schltr.	S	Talbot 3724
B.schinzianum Kränzl.	N	Barter 1482
B.barbigerum Lindl.	S	Holland 297
B.fuscum Lindl.	N, S	King 191, Meikle 599
B.oreonastes Rchb.f.	N, S	Hepper 1430, Richards 3457
B.zenkeranum Kränzl.	N	Wimbush in King 135
B.nigericum Summerh.	N	King 124
B.lupulinum Lindl.	N, S	Latio & Daramola FHI 28943, Brenan *et al.* 8643
B.porphyrostachys Summerh.	S	Keay FHI 22711
B.falcipetalum Lindl.	S	Mann 526
B.rhizophorae Lindl.	S	Talbot s.n.
B.congolanum Schltr.	N, S	King 21, Keay FHI 25426
B.fuerstenbergianum (De Wild.) De Wild.	N, S	Wimbush in King 129, Cooper 38
B.tentaculigerum Rchb.f.	S	Latilo FHI 41331
B.simonii Summerh.	S	Simon 11
B.melanorrhachis (Rchb.f.) Rchb.f. ex De Wild.	S	Millen s.n.
B.falcatum (Lindl.) Rchb.f.	S	Brenan FHI 22713
B.bufo (Lindl.) Rchb.f.	S	Keay FHI 25474
B.calyptratum Kränzl.	S	Brenan FHI 22735
B.colubrinum (Rchb.f.) Rchb.f.	S	Keay FHI 28129
B.oxypterum (Lindl.) Rchb.f.	N, S	King 126, Wright 134
B.maximum (Lindl.) Rchb.f.	S	Barter 1854
B.kamerunense Schltr.	S	Sanford 136/66
B.magnibracteatum Summerh.	S	Sanford 1726/65
B.imbricatum Lindl.	S	Wright 149
B.leucorrhachis (Rolfe) Schltr.	S	Millen 189
Genyorchis apertiflora Summerh.	S	Keay FHI 25545
G.pumila (Sw.) Schltr.	S	Barter 72
Oeceoclades maculata Lindl.	N, S	Onochie FHI 18259, Thomas 2196
O.saundersiana (Rchb.f.) Garay & P.Taylor	S	Burtt B28
Eulophia guineensis Lindl.	N, S	Parsons L106, Kennedy 2651
E.quartiniana A.Rich.	N, S	Barter 1485, Keay FHI 16210

Species	Distribution in Nigeria	Reference herbarium specimen (K)
E.gracilis Lindl.	N, S	Westwood 276, Dalz. 1284
E.euglossa (Rchb.f.) Rchb.f.	N, S	Ajayi FHI 19300, Talbot 162
E.stachyodes Rchb.f.	N	King 119
E.zeyheri Hook.f.	N	King 155
E.adenoglossa (Lindl.) Rchb.f.	N	Dalziel 562
E.shupangae (Rchb.f.) Kränzl.	N, S	Lely 353, Rosevear 67/29
E.lindiana Kränzl.	N	G.V.Summerhayes 51
E.milnei Rchb.f.	S	Robb s.n.
E.warneckeana Kränzl.	N, S	Dalziel 446, Sanford WS/899/65
E.monile Rchb.f.	N	Hepburn 55
E.brevipetala Rolfe	N	E.W.Jones 139
E.ramifera Summerh.	N	Hepburn 113
E.sordida Kränzl.	N, S	King 61, Richards 5026
E.parvula (Rendle) Summerh.	N	G.V.Summerhayes 142, 143
E.leonensis Rolfe	N	Keay FHI 25777
E.galeoloides Kränzl.	S	Tamajong FHI 20295
E.alta (Linn.) Fawcett & Rendle	S	Millen s.n.
E.flavopurpurea (Rchb.f.) Rolfe	N, S	Millson 86, Foster 209
E.cristata (Sw.) Steud.	N, S	Dalziel 287, Burtt B33
E.caricifolia (Rchb.f.) Summerh.	S	Dalziel 964
E.angolensis (Rchb.f.) Summerh.	N, S	Barter 1486, Richards 5087
E.cucullata (Sw.) Steud.	N, S	Barter 1488, Punch s.n.
E.dilecta (Rchb.f.) Schltr.	N, S	Westwood s.n., Talbot s.n.
E.buettneri (Kränzl.) Summerh.	N, S	King 195, Keay FHI 23436
E.horsfallii (Batem.) Summerh.	N, S	Parsons L107, Phillips 46
E.barteri Summerh.	N	Barter 1481
E.oedoplectron Summerh.	S	Shepherd 6
E.latilabris Summerh.	N	G.V.Summerhayes 118
E.orthoplectra (Rchb.f.) Summerh.	N, S	McClintock 207, Maitland 1778
Graphorkis lurida (Sw.) O.Ktze.	N, S	King 40a, Clarke 28
Podangis dactyloceras (Rchb.f.) Schltr.	N	G.V.Summerhayes 140
Calyptrochilum emarginatum (Sw.) Schltr.	S	Ross 246
C.christyanum (Rchb.f.) Summerh.	N, S	Keay & Mutch FHI 22933, Keay FHI 22487
Angraecum distichum Lindl.	N, S	King 105a, Meikle 519
A.aporoides Summerh.	S	Barter 1854
A.podochiloides Schltr.	S	Talbot 894
A.pungens Schltr.	S	Talbot 892
A.subulatum Lindl.	S	Kennedy 1796
A.multinominatum Rendle	S	Barter 3352
A.pyriforme Summerh.	S	Richards 3476
A.angustum (Rolfe) Summerh.	S	Holland 27
A.chevalieri Summerh.	N, S	King 113, Jones FHI 20223
A.egertonii Rendle	S	Wright 157
A.angustipetalum Rendle	S	Kennedy 1918
A.birrimense Rolfe	S	Kennedy 280

Species	Distribution in Nigeria	Reference herbarium specimen (K)
A. eichleranum Kränzl.	S	Talbot 899
A. infundibulare Lindl.	S	Barter 1860
Listrostachys pertusa (Lindl.) Rchb.f.	S	Mann 524
Microcoelia guyoniana (Rchb.f.) Summerh.	N	King s.n.
M. microglossa Summerh.	S	Talbot s.n.
M. caespitosa (Rolfe) Summerh.	S	Sanford WS/744/64
M. dahomeensis (Finet) Summerh.	S	Sanford WS/164/66
M. koehleri (Schltr.) Summerh.	S	Keay FHI 22430
Encheiridion macrorrhynchium (Schltr.) Summerh.	S, N	Daramola & Adebusuyi FHI 38404, Sanford 1356/65
Chauliodon buntingii Summerh.	S	Maggs 125
Diaphananthe dorotheae (Rendle) Summerh.	S	Talbot 914
D. pellucida (Lindl.) Schltr.	S	Jones FHI 19536
D. plehniana (Schltr.) Schltr.	S	Sanford 121/66
D. bidens (Sw.) Schltr.	S	Ross 283
D. curvata (Rolfe) Summerh.	S	Keay FHI 16243
D. laticalcar J.B.Hall	S	Sanford WS 4810
D. longicalcar (Summerh.) Summerh.	S	Talbot 939
D. rutila (Rchb.f.) Summerh.	N, S	King 125, Charter FHI 38714
D. obanensis (Rendle) Summerh.	S	Talbot 896
Bolusiella talbotii (Rendle) Summerh.	S	Talbot 941
Chamaeangis ichneumonea (Lindl.) Schltr.	S	Keay FHI 22704
C. lanceolata Summerh.	S	Markham FHI 50083
C. odoratissima (Rchb.f.) Schltr.	N, S	King 190, Floyer 6
C. vesicata (Lindl.) Schltr.	S	Keay FHI 26733
C. sp. *nov.*	S	Segerbäck 1169
Plectrelminthus caudatus (Lindl.) Summerh.	N, S	Onochie FHI 18265, Barter 1858
Aërangis kotschyana (Rchb.f.) Schltr.	N	Hepburn 82
A. biloba (Lindl.) Schltr.	N, S	Onochie FHI 35947, Keay FHI 22432
Rangaëris muscicola (Rchb.f.) Summerh.	S	Keay FHI 25380
R. rhipsalisocia (Rchb.f.) Summerh.	S	Carpenter 177
R. longicaudata (Rolfe) Summerh.	S	Cooper 79
R. trilobata Summerh.	S	Talbot 3299
R. brachyceras (Summerh.) Summerh.	S	Cooper 42
Cyrtorchis ringens (Rchb.f.) Summerh.	S	Moloney s.n.
C. monteiroae (Rchb.f.) Schltr.	S	Talbot 936
C. chailluana (Hook.f.) Schltr.	S	Mann 521
C. arcuata (Lindl.) Schltr. subsp. *variabilis* Summerh.	N, S	Keay FHI 22899, Kitson s.n.
C. hamata (Rolfe) Schltr.	N, S	Westwood 270, Keay FHI 25373
C. aschersonii (Kränzl.) Schltr.	S	Kennedy 234
Solenangis scandens (Schltr.) Schltr.	S	Onochie FHI 23423
S. clavata (Rolfe) Schltr.	S	Moloney s.n.
Dinklageëlla liberica Mansf.	S	Kennedy 1900

Species	Distribution in Nigeria	Reference herbarium specimen (K)
Eurychone rothschildiana (O'Brien) Schltr.	N, S	King 196, Wright 155
Ancistrorhynchus clandestinus (Lindl.) Schltr.	N, S	King s.n., Sanford FHI 56582
A.recurvus Finet	S	Wright 158
A.capitatus (Lindl.) Summerh.	S	Barter 1857
A.cephalotes (Rchb.f.) Summerh.	N, S	G.V.Summerhayes 85, Sanford WS/1680/65
A.metteniae (Kränzl.) Summerh.	N, S	Cole 10, Moloney s.n.
A.serratus Summerh.	S	Head 95
A.straussii (Schltr.) Schltr.	S	Wright 103
A.schumannii (Kränzl.) Summerh.	S	Talbot 895
Angraecopsis ischnopus (Schltr.) Schltr.	N	Wimbush in King 132
A.elliptica Summerh.	N	King 57
Tridactyle anthomaniaca (Rchb.f.) Summerh.	S S	Ross 228
T.muriculata (Rendle) Schltr.	S	Talbot 904
T.oblongifolia Summerh.	S	F.H.I. Staff FHI 42061
T.lagosensis (Rolfe) Schltr.	S	Barter s.n.
T.tridactylites (Rolfe) Schltr.	N	Onochie FHI 35380
T.bicaudata (Lindl.) Schltr.	N	King 15
T.gentilii (De Wild.) Schltr.	N	King 192

FHI = Forest Herbarium Ibadan

Bibliography

Hutchinson, J. & J.M.Dalziel 1968. *Flora of West Tropical Africa,* vol.III, Part 1. Second Edition, revision edited by F.N.Hepper. Crown Agents, London.

Morton, J.K. 1961. *West African Lilies and Orchids* (West African Nature Handbooks). Longmans, London.

Piers, F. 1968. *Orchids of East Africa.* Verlag von J.Cramer, Lehre.

Stewart, J. & B.Campbell 1970. *Orchids of Tropical Africa.* W.H.Allen, London.

Summerhayes, V.S. 1968. Orchidaceae (Part 1). In: E.Milne-Redhead & R.M.Polhill (eds), *Flora of Tropical East Africa.* Crown Agents, London.

Williamson, G. 1977. *The Orchids of South Central Africa.* J.M.Dent & Sons Ltd, London.

Index to plants mentioned in the text